Gardening Indoors with Cuttings

by George F. Van Patten

Alyssa F. Bust

Published by Van Patten Publishing
Cover Design: Tin Man Design
Artwork: Hart James
Book Design: G. F. Van Patten
Cover Photo: Courtesy Gene Klump (*Mr. Vegetable*) - Taken in the Garden Shop & Nursery, Reno, Nevada.
Back Cover Photos Courtesy Green Air Products

ISBN 1-878823-20-5
First Printing
9 8 7 6 5 4 3 2 1

This book is dedicated to the editors that made *Gardening Indoors with Cuttings* the best book possible!
Thank you editors!!

Australia - Vivien Ireland, Robin Moseby, Graeme Plummer, Barry Silver

Canada - Joen Belanger, Andre Courte, Guy Dionne, Tom Duncan, Scott Hammond, Sharon Harper, Keith Harper, R. Markiewicz, Frank Pastor, Russ Rea, Shelly Rea, Bob Rogers, Don Stewart, William Sutherland, Francois Wolf, Jr., Wayne Hayes, Ron Hayes

New Zealand - Rob Smith

UK - Giles Gunstone

USA - Tom Alexander, Carl Anderson, Russ Antkowiak, Sarah Bares, C."Gar" Bares, Allan Bednar, Brandy Bolt, Teg Bradley, Larry Brooke, Sunny Bueck, Brian Bueck, Michael Christian, Mosen Daha, Peter DePaola, Vince DInapoli, Bob Edberg, Jeff Edwards, Bill Fetner Jr., Ed German, Jeff Gibson, Chris Guthrie, Kim Hanna, Wayne Hayes, Martin Heydt, Jim Howell, Christine Hubbard, David Ittel, Darryl Johnston, Patrick Joyce, Ron Kleinman, Rick Martin, Tiffany Martin, Richard Middlebrook, Richard Miller, Dorothy Morgan, Tim Olivia, Tracy Peltzer, John Pierce, Nancy Pierce, Waymon Price, Brandy Price, Rajaim Purcifel, Harmon Davidson, Bill Ross, Chris Rothe, Chris Schneider, Tom Shelsky, Steve Stragnola, Brooke Taggert, Roger Thayer, Evan Townsley, Richey Truce, Jerry Van Volkenburg, Patrick Vivian, Peter Wardenberg

Table of Contents

Introduction

Successful plant propagation depends on environmental conditions. Gardeners must supply proper lighting, humidity, air and root-zone heat, growing medium and rooting hormones. In addition, gardeners must be wise when selecting a mother plant and choosing a method of taking cuttings. In this book, we examine these issues in order to give the reader a full understanding of the process of plant propagation. This book is a helpful resource for hobbyists beginning or continuing to learn about the myriad factors involved in propagating plants asexually.

Plant propagation refers to the methods used to increase plant stock. Clearly plants propagate themselves in their natural surroundings. As a child perhaps you blew the seeds from a dandelion stem and increased the yellow patches on your family's lawn. Or perhaps you have pulled weeds in your garden only to find them reappear several weeks later. These common examples illustrate the two different means by which plants propagate themselves: sexually and asexually.

A plant seed is the by-product of sexual reproduction and contains all the genetic characteristics of a plant. A seed has both male and female genes from its parents. Most plant seeds need only water, heat and air to germinate. After a seedling is established, it enters the vegetative growth stage. During the vegetative stage plants produce green leafy growth. When a plant is producing an optimum amount of chlorophyll, it will grow as much as its environment will allow. For optimum growth, a gardener must supply ideal levels of light, heat, water, carbon dioxide, fresh air, and nutrients.

During a plant's vegetative growth, it can be asexually reproduced. Cuttings are the result of asexual or vegetative propagation. Taking a cutting is simply cutting a growing piece of a plant and inducing it to establish its own rooting system. For best results, cuttings are taken from a mature mother plant in its vegetative stage. It is possible to take a cutting from a flowering plant, but it is likely to grow more slowly.

Although both sexual and asexual propagation are easy and fun, this book focuses on the asexual methods. Why are plants able to reproduce themselves asexually? It seems almost mystical that leaf or stem cuttings can grow root systems of their own. All cuttings, regardless of where on the plant they come from, contain natural rooting hormones known as auxins. These natural hormones, which are often supplemented by commercial rooting hormones, are growth stimulants which activate the rooting process.

One of the primary benefits of propagating from cuttings is the relative predictability you enjoy. When you grow from seeds, you are not entirely sure of the end-result. However, when you take a plant cutting, the new

Taking a cutting from a mother plant.

plant will be an exact genetic replica of the mother plant. This has advantages and disadvantages. If your mother plant is weak and sickly, your cutting will be as well. However, if you take a cutting from a strong, succulent plant, follow some simple procedures outlined in this book, and provide a good growing environment, you will enjoy another strong and succulent plant. This allows you to maintain consistency and uniformity amongst your plant stock.

Another benefit of asexual propagation is the flexibility it provides for disease control. Because plants taken from cuttings tend to be stronger and larger for their age than their counterparts grown from seed, they are more disease-resistant. If you take a cutting from a healthy, disease-resistant plant, you produce a disease-resistant cutting. If an insect infestation affects your crop, it is often easy to eliminate by throwing away any diseased cuttings, removing the cuttings from the garden room (this is easy because many propagation trays hold up to one-hundred cuttings) and fumigating. With careful selection followed by consistent aftercare, many gardeners boast a 100% yield from cuttings.

Finally, cuttings are superior to seeds if you are interested in performing experiments. Normally cuttings taken from the same mother plant will be identical. However, cuttings will differ if they are grown in different environments. You can easily examine and record the effects of fertilizers, light, water, and rooting hormones. If you introduce various stimuli to a select group of cuttings all taken from the same parent plant, you will have the ability to make a true comparative analysis and determine ideal growing conditions for your cuttings. Cuttings grown in a poor environment will be weak and perhaps stunted. Cuttings grown is an optimum environment will be strong. It is worth your time to keep good records so you can maintain and reproduce environments which help to produce healthy cuttings.

Rule of Thumb: Cuttings should root in three weeks or less.

Many parts of a plant can be used for propagation. You can root leaves, stems, growing tips, or buds. Although many parts of a plant will root, they are not all equal in terms of time required, ease of rooting, attention demanded, and equipment needed. For instance, leaves will root, but generally take twice as long as growing tips. Because different methods of rooting cuttings require vari-

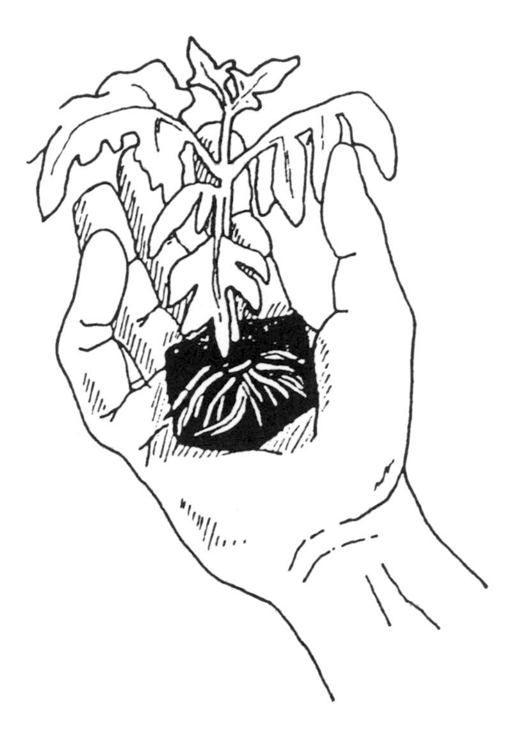

Rooted cutting in rockwool cube.

ous degrees of skill, time, and equipment (which often equals money), in this book we will focus on a simple means of propagation using stem cuttings. This method requires the least amount of sophisticated equipment and produces quick results. Often stem cuttings will produce roots in 10 days or less. Even as a beginner, you are virtually guaranteed to have success with this method.

Once you have stem cuttings mastered, you might want to experiment. Provided that you pay attention to the basics– lighting, humidity, temperature, rooting medium, and rooting hormones– you should be able to successfully propagate plants using a variety of techniques. This book discusses some alternatives that you may wish to try: leaf or root cuttings, simple layering, air layering, and grafting.

For the indoor gardener, the benefits of propagating your own plants are countless. It is an economical way to increase your plant collection. Once you have the basic equipment to provide an adequate environment, propagation is virtually free and certainly less expensive than buying your plants from a nursery. You can gather cuttings from friends, neighbors, public gardens, and local parks. Public gardens and parks often have unusual varieties of plants which would make welcome additions to most gardeners' collections. Be sure to ask an authority before taking a cutting. If you collect native plants on hikes, to the best of your ability try to recreate the environment where the native flora grew. The more similar the environment you provide is to the natural one, the more success you will have. For instance, a cutting from a plant which is accustomed to growing in rock and gravel on a steep slope of a mountain side is not likely to thrive in very wet soil.

This chapter discusses lighting, humidity, and temperature. In order to create a healthy growing environment for your clones, it is important to understand how these three environmental components affect your cuttings. Remember, when you grow plants indoors you are responsible for playing Mother Nature. Providing optimum light, humidity, and heat for your cuttings is one way to help ensure success.

Chapter One
The Environment

Lighting

Light is important to plant growth because it is one of the primary ingredients in a plant's recipe for food. A plant combines light energy with carbon dioxide, water, and nutrients, to produce chlorophyll and carbohydrates. Without light, a plant will turn yellow (and eventually die) because it can not produce green chlorophyll.

When thinking about lighting for your cuttings, you must consider two aspects: the amount and the type of light you will use. Research suggests, and gardeners confirm, that lighting greatly effects how quickly and successfully cuttings root.

You can obtain specific information about the amount of light various plants need from the Department of Agriculture. The information will specify how much light is needed for different stages of a plant's life. Remember, there is a difference between the amount of light emitted from a source and the amount of light received by the plant. Although many lights are rated in watts-per-square-foot, the most accurate way to measure the intensity of the light that plants receive is using foot-candles. A foot-candle is the amount of light that falls on a surface, equal to the one lumen per square foot. Light meters measure foot-candles and are available at most garden supply or camera stores. This inexpensive investment will ensure that you get the most out of your lights. In general, studies have shown that the optimum light level for rooting leafy cuttings falls between 200 and 1,000 foot-candles (between 20 and 100 watts-per-square-meter). This translates into between 50 and 95 percent shade on a sunny day.

Most gardeners provide 18 hours of light per day for their cuttings. There are some gardeners who prefer 24 hours of light, saying the cuttings respond faster. You might need to experiment to find out what works best for you. In determining how long a light period to maintain, remember that your cuttings have been conditioned by the mother plant and are accustomed to her environment. If the mother plant received 18 hours of light a day, that is what the cutting will be preconditioned to expect. Through auxins and hormones, the mother plant has predetermined the success of the cutting based on her environmental influences.

The only time it is beneficial to deviate from the lighting schedule provided for the mother plant is if you are interested in making your cutting a better phenotype than the mother plant. Factors that are inherent to a plant because of genes are referred to as the genotype. The phenotype, on the other hand, is the outcome of environmental factors. It determines the plant's ability to survive. Often the potential supplied by the genotype is not realized. If your mother plant is a good genotype, but a poor phenotype, you might want to increase lighting time to prod the cutting to be a better phenotype. In general, however, gardeners achieved the most success by matching the photoperiod of the mother plant, often through 18 hours of light.

When choosing a light source, you have several options– natural light, fluorescent lights, and high intensity discharge (HID) lamps. If you choose to grow your cuttings outdoors, you can rely on natural lighting. Be sure to provide some shade by draping a cover over your growing area or by placing your cuttings by the base of a tree where they will be sheltered and shaded by branches. Without being filtered through clouds, dust, pollution, or vegetation, full sunlight has the potential of being up to twice the light saturation level for cuttings. *Shading is imperative!*

For the purpose of this book, however, we are focusing on growing cuttings indoors, in which case natural light is not really a viable option. If you live in a particularly sunny location, and can place your propagation trays in an east or west facing window that is light, but that does not receive full direct sunlight, you might have some success. However, because cuttings thrive on 18 to 24 hours of light a day, it is unlikely that natural lighting will be sufficient without supplemental lighting.

For more predictable results, most growers choose fluorescent lights. Because fluorescents produce very little heat they can be placed within inches of propagation trays. A 4-foot lighting fixture with two 40-watt fluorescent tubes will support two trays of up to one hundred cuttings each. For more light in the blue spectrum (plants use light primarily from the red to blue spectrums), look for bulbs with a higher Kelvin temperature. You can easily find bulbs with a Kelvin output of more than 5000° in garden stores. For bulbs with a Kelvin output of more than 6,000^{o} you might have to rely on scientific supply

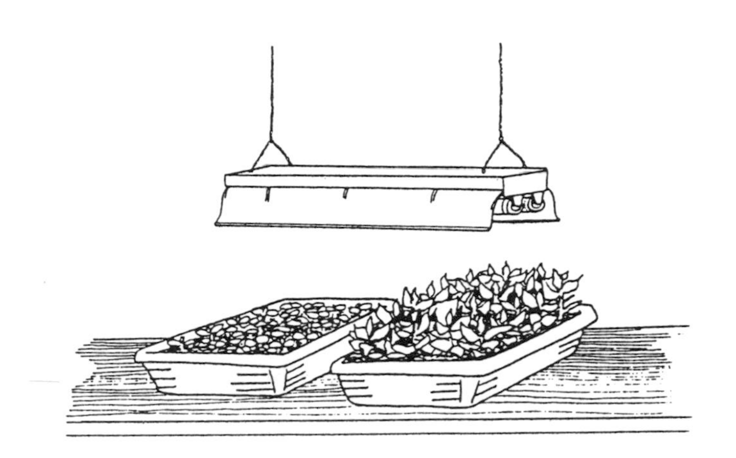

Fluorescent lights are the most commonly used source of light to root cuttings. Fluorescents provide low levels of light cuttings need during rooting.

houses. When shopping for a fluorescent light, look for a fixture which houses two tubes and a reflective hood.

Kelvin is the absolute temperature of an object. 0° Kelvin = -273.6 Celsius. 0° Kelvin is absolute zero. It can not get any colder. The visible spectrum of light is red (low energy photon) at about $3{,}000^{o}$ Kelvin to green ($5{,}500^{o}$ Kelvin) to high-energy blue (5,500 to $6{,}000^{o}$ Kelvin). The higher the Kelvin temperature, the shorter the wavelength and the higher the frequency of the light photon.

Many fluorescent bulbs work well to root cuttings including: warm white/cool white (one each) PowerTwist, Growlux, 5000 KV (daylight) bulbs, etc. For a complete rundown on the best bulbs, ballasts and reflectors to use to get the most out of your fluorescent bulbs, refer to *Gardening Indoors with Fluorescent Lights,* by George F. Van Patten and Alyssa F. Bust.

Because HID lamps generate a lot of heat, they are

A single halide lamp can be hung at least 3 feet above several flats of cuttings. It is easy to water the cuttings with a water wand.

slightly trickier to use. You need a space where you can suspend the lights three to four feet above the cuttings. If you use HID lamps, you should reduce light intensity and prevent additional stress on your cuttings by shading them. Until cuttings have rooted, they rely on their leaves for transpiration. Shade helps keep leaves from drying out so that the cutting can continue to manufacture carbohydrates. Covering your propagation box with cheesecloth, white-washed glass, netting, or plastic or cloth screening will provide adequate shade. Once roots are established, some of the stress is alleviated and shade can be decreased. At this point, many gardeners choose to increase light levels to help stimulate root growth. Place your lights as close to your plants as possible without burning them. This could be as close as two to four inches, but use caution. HID lamps are much hotter and must be farther away, 18-36 inches. Once rooted, the more light plants receive, the faster they will grow.

Make sure to use a grounded light system. Remember water and electricity don't mix. A grounded circuit could save your life!

Light is not the only ingredient in a plant's recipe for making food. Carbon dioxide (CO_2) is also a vital component. Add CO_2 to boost the cuttings' growth. For more information on exactly how CO_2 affects plant growth refer to *Gardening Indoors with CO_2* by George F. Van Patten, Alyssa F. Bust and Tom LaSpina. Remember, light is a vital energy supplier for the metabolic processes that enable rooting. If your light levels fall below optimum, it is possible that the quantity of carbohydrates produced by photosynthesis will be limited, subsequently decreasing rooting.

Humidity

Providing an adequate water supply and humidity level is perhaps the largest obstacle beginning (and experienced) gardeners face. If your lights are too intense or too close to your cuttings before they have established roots, or if your garden environment is not humid enough, your cuttings are likely to dry out. If, on the other hand, you provide too much water, you can promote root rot. Nurseries across the country report that root rot is one of the most common concerns their customers express when they are learning about cuttings. You want to keep your growing medium moist, but not soggy. Overwatering can cause damage and potentially death, to your vulnerable cuttings. In general you will need to lightly water your cuttings every two to three days. One way to ensure that all the cuttings are watered equally is to water from one end of the tray while gently tipping it upwards, forcing the water to the other end. Repeat from the other side before laying the tray flat again. Another way is to fill a five-gallon bucket with water. Put a pump on a timer in the bucket. Set the timer to water the cuttings daily.

While you want to keep your growing medium only moist, you should keep the air surrounding the cuttings humid. Try to maintain ninety percent humidity in your propagation area. Without roots, a plant must rely on its leaves and stem to take in nutrients. This is why cuttings taken with leaves are often easier to propagate than those without. Cuttings with leaves can continue to produce carbohydrates if they are provided with proper growing conditions.

Rule of Thumb: Keep the propagation area at or near 90 percent humidity.

Correct humidity is an important environmental concern because it determines a plant's rate of transpiration. Because the relative humidity inside a leaf is generally 100 % while the surrounding air is less humid, the leaf looses water to the air through diffusion and evaporation. With increased transpiration, the cutting experiences water stress and it uses stored food (carbohydrates) for activities other than root production. When a cutting experiences water stress, stomata close in an attempt to reduce water loss. When the stomata close, carbon dioxide is unable to enter the plant's system and growth slows. Carbon dioxide is one of the crucial ingredients of photosynthesis; without it plant growth slows to a halt. When roots form, water stress is alleviated, stomata open, and photosynthesis resumes.

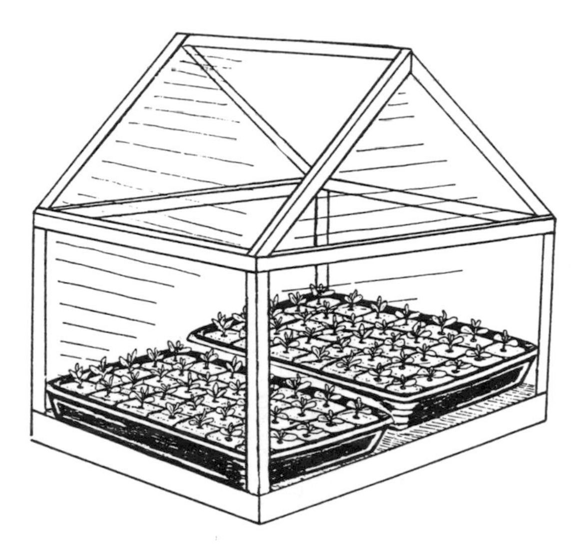

This humidity tent is made from a few boards framed together. Clear plastic is stretched over the top and a gap is left around the bottom to allow for air circulation.

There are ways to reduce the stress before roots form. By providing humid air, water stress is reduced and photosynthesis is increased. The more humid the environment, the less transpiration and water stress, and the more root growth your plant will experience.

There are several easy ways to increase and maintain proper humidity levels. Cloning wax, one of many antitranspirants, contains nutrients and hormones and can be sprayed onto your cuttings. Antitranspirants work by clogging a plant's stomata, thereby preventing water loss through transpiration. There is a drawback to this method. Closed stomata block water loss and carbon dioxide exchange. Without adequate carbon dioxide, photosynthesis is slowed. If you must use antitranspirants, stop using them as soon as cuttings are rooted.

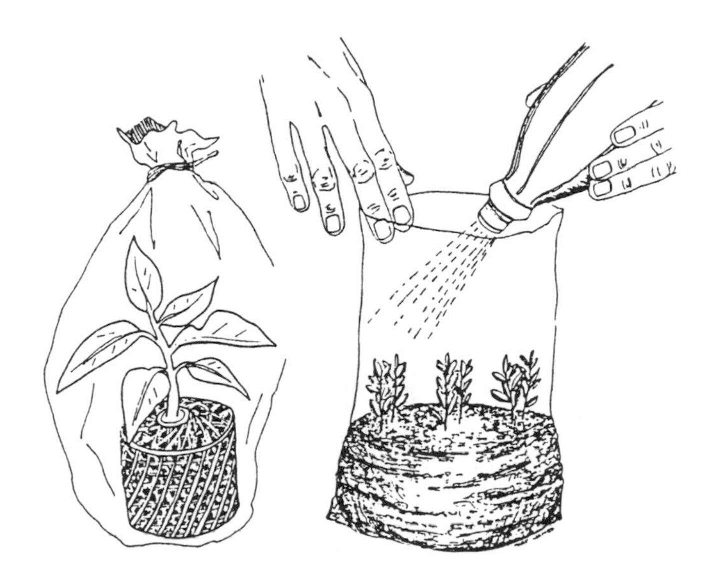

A plastic bag can make a perfect humidity tent for a few plants.

Misting your cuttings with warm water or solution is the least complicated way to create humidity. If your cuttings are in a contained garden room, you can maintain high humidity by wetting not only your cuttings, but also the walls and the floor of your room. A syringe works well to accomplish this. Most gardeners, however, choose a method which is ultimately less labor intensive and more fool-proof.

Caution! Be careful to keep water off low light bulbs when misting. The water/temperature stress on the bulb could cause it to break.

Creating an environment which is fully enclosed can increase your humidity level and therefore your success rate. The most common way to achieve this is to buy or make a humidity dome. Garden suppliers sell inexpensive plastic covers that fit propagation trays. You can fabricate your own dome by using plastic and wire or sticks which serve as "tent poles" to keep the plastic off the plants. Simply put some sticks in your growing medium and cover your propagation tray with plastic. If you are propagating directly into pots, create a tent over each one. This method is clearly time consuming, but it is the trade-off for not having to replant your cuttings once they have rooted.

Some gardeners have discovered that clear plastic sweater boxes work well for propagation. They can be purchased in a variety of sizes and come with a lid that helps promote humidity. Remember, sterilization is key. Sterilize the box before using it with a mild bleach solution. Fish tanks also work well. Sterilize the terrarium, and cover it with plastic if it does not have its own glass cover. A twenty gallon tank can comfortably hold 35 cuttings.

If you do not want to create an enclosure, you can simply lay some damp cloth or paper directly on top of the cuttings. This will keep the air immediately available to the cuttings moist. Although this is an easy, inexpensive, and effective method, there are two major drawbacks. Direct-contact limits both air circulation (discussed below) and light. If light is reduced too drastically, photosynthesis potential decreases and growth may be slowed. Direct contact also promotes bacterial and fungal infections.

Humidity domes, however you choose to obtain one, do not produce humidity. They simply contain it. You must mist your cuttings with warm, preferably distilled, water every two days to ensure that there is enough moisture in the dome.

If you want a step up in sophistication from the plastic-over-the-pot method, several home-made and sev-

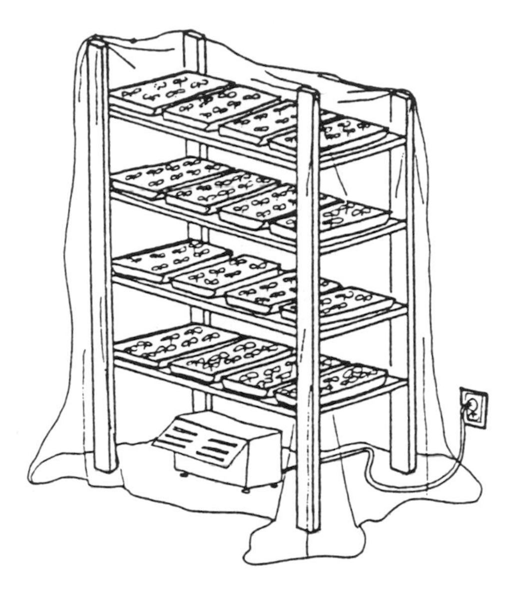

This humidity tent was easy to make. The gardener used an old shower curtain to drape over shelf holding cuttings. Note the heater below.

eral more expensive and elaborate store-bought alternatives are available. You can set up a misting box by venting a humidifier into an enclosed propagation area. Simply construct a frame from a durable material. Plastic pipes work well, as does wood which has been treated for moisture-resistance. The frame should form a simple box outline. The size depends on your needs. Cover the frame with plastic. With clear plastic packaging tape, join one end of the plastic covering to the mouth of the humidifier. When the humidifier is on, you will have an enclosed, high-humidity, growing area. You can suspend a light above your set-up and attach both the light and the humidifier to timers and even a humidistat if you want to become automated. For best results, run the humidifier for 30 minutes every two hours, or in 15 minute spurts every hour.

It is also possible to set up an automatic misting box which contains a pipe and a mist nozzle. Simply run a pipe from a garden hose (you will need a connector to fit the pipe and the garden hose together) into the center of your misting box. You will use two pipes, one horizontal and one vertical. The horizontal pipe supplies water from a garden hose into the box. This will require either a hole in the bottom of the box or an opening at the base of one of the sides. To the horizontal pipe, attach a vertical one with a mist nozzle at the top, ideally about 18 inches high. Attach the garden hose to a faucet. Turn the faucet on slightly and you will have an automatic mister which will keep leaves and stems moist while your cuttings are forming their roots. The mister can be left on throughout the day. As a general rule, use it during the warmth of the day, not at night. If you want to get high-tech, you can even attach solenoids valves and an electric timer so you can mist intermittently, like many large-scale production propagators.

If you choose to generate humidity by providing an enclosed environment or through direct contact, do not forget that plants require air circulation. Once a day, take the dome off your tray and shake off any condensation which has developed. If you are faithfully doing this and your plants still show signs of wilting, cut holes in your humidity dome/cover to allow for further air circulation. Some gardeners recommend leaving 3 to 4 inches of a humidity box uncovered to allow for air circulation. Humidity is essential, but without air circulation your cuttings may be subject to damping off disease or mold.

For growers with larger operations, a spinning-impeller disk and cool-mist humidifier can raise the humidity level of your garden room to an optimum 90

This humidity tent is made from plexiglass. Holes were drilled in the sides to allow air to circulate.

percent. When purchasing a humidifier, be sure not to buy an ultrasonic or conventional steam system. They may add humidity to your room, but they may also add oxides of heavy metals. Fill your humidifier with water that is disinfected. A mixture of one teaspoon of household bleach per gallon of water will help prevent algae and bacteria.

Large propagation facilities also make use of fogging, a system which requires less water than misting and therefore reduces chances of over-watering. Fogging systems distribute very small water droplets and maintain a consistently high humidity level. This is achieved by forcing water at a very high pressure through a tiny nozzle aperture. Because the water droplets are so small, they remain air-borne until the air is saturated. One commer-

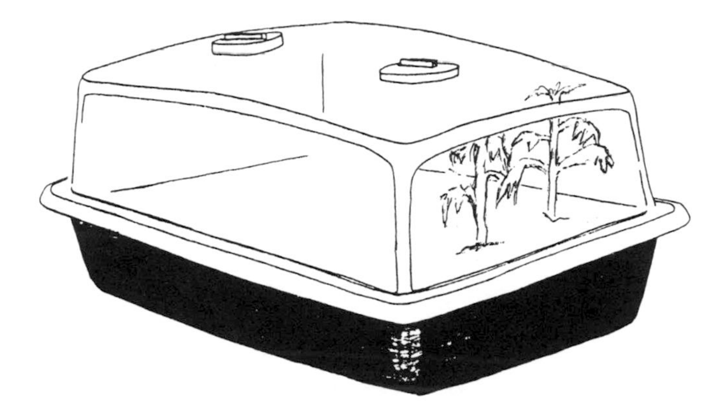

Vents can be opened or closed in this humidity tent to control air circulation.

cial fogging system vibrates water on a ceramic plate at a high frequency so the water vaporizes. Misting, however, produces water droplets that range in size, but are considerably larger than those produced by a fogging system. Because they are large, these water droplets fall to the plants quickly. Fogging systems are expensive and have limited applications for most home hobbyists.

Whatever method you choose in order to maintain humidity, remember that failure to maintain humidity during the first two weeks of rooting is likely to cause your cuttings to die. Be diligent about misting to ensure proper humidity. Once your cuttings have rooted, increase the amount of time you leave the cover off your misting box, until finally it is off all the time. Once roots are established, water stress is not such a concern and it is important to provide air circulation so your plants will have access to fresh carbon dioxide-rich air.

This aeroponic cutting chamber has lots of holes drilled in the top with a misting nozzle system set up in a big black box. The cuttings hang down into the big black box and get misted. The mist water sets in the bottom of the tub and is pumped up every 1-5 minutes..

One little trick to make slow-rooting plants root is to break the one-inch rockwool cube apart to give it extra air. The plant should root in 1-3 days.

Temperature

Indoor gardeners must think about regulating the temperature of both the air and the rooting medium in which their cuttings are planted. Temperature directly affects photosynthesis, respiration, and therefore root growth. Excessively low temperatures can disrupt a plant's metabolic processes and slow plant growth.

Providing bottom heat for cuttings is one of the

easiest ways to increase your success rate. Simply use a heating mat underneath your propagation tray to provide warmth to the rooting medium and the roots themselves. Propagation mats are available at many garden stores. Many have built-in thermostats. Those that don't can be used with an adjustable thermostat that you set yourself.

If you are only rooting a couple of cuttings, you can cover an unshaded light bulb with an inverted pan and place you propagation tray or pots on top of the pan. The light bulb will provide a gentle heat to the pan. Be sure to monitor this set-up carefully.

Rule of Thumb: Maintain the bottom heat temperature between 78-80 degrees F. both day and night.

Rule of Thumb: Keep the air 5-10 degrees cooler than the rooting medium, generally between 65-75 degrees F. both day and night.

In order to regulate heat, place a thermometer in the rooting medium. If you are growing a lot of clones, you might want to scatter thermometers around your garden area to ensure even heat distribution. If you are using a common heating pad, or if your propagation mat does not have a built-in thermostat, you may have to experiment with your heating device until you can determine the proper setting. Heating pads, for instance, can not be set to a specific temperature and a "low" setting on one heating pad is likely to be different to the "low" setting on another unit. Use a thermometer in the rooting medium to determine the temperature. Ideally, you want to maintain your bottom heat temperature between 78 and 80 degrees. If the pad is too hot, you can place a spacer between the pad and the cuttings to lower the temperature. After roots are established, lower root-zone tempera-

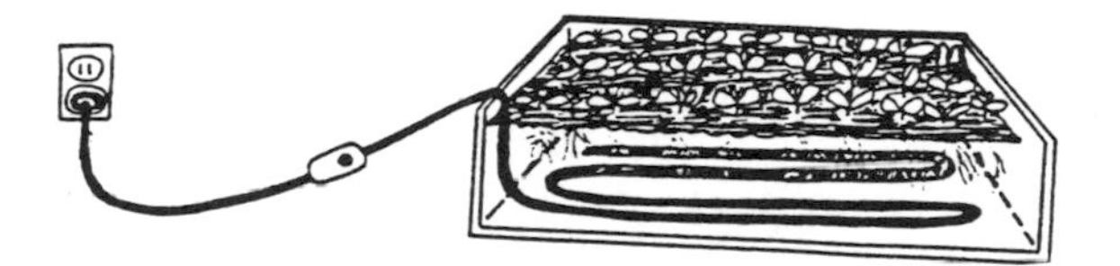

Heat cables or a heating pad provide bottom heat which stimulates faster root growth.

tures slightly. Check the thermometer often to ensure that proper temperatures are maintained. Many gardeners maintain that correct bottom heat temperature is the most important factor for successful and fast rooting.

If you decide that propagating is a passion and you want to invest in some good equipment, propagation benches like those used by large greenhouses are available for home gardeners. If you need to heat a propagation area between 100 and 550 square feet, you can purchase a warming bench which regulates bottom heat. Although they add ease to the busy schedules of large-scale growers, for the hobbyist, they are expensive and do not ensure any better results than the home-made alternatives discussed above.

Air temperatures should be slightly cooler than root-zone temperatures. This will help slow transpiration. Misting and humidity help maintain cooler air tempera-

An incandescent bulb controlled by a rheostat is an excellent way to control bottom heat.

tures. Keep the air five to ten degrees cooler than the rooting medium, generally between 65 and 75. You might need to use a space heater to maintain the heat at night. The climate inside the humidity box is easiest to control when the temperature outside the box is between fifty and eighty degrees. This rules out unheated garages for winter propagation.

Chapter Two
Nutrient Solutions

Nutrient Solutions

Cuttings need all the help they can get to root into healthy plants. Remember, the quicker your cuttings grow roots, the less stress they will endure. Vitamin B^1 helps alleviate the shock cuttings experience. It is available in liquid form at garden stores. You can also help by feeding your cuttings a nutrient solution to provide them with as many micronutrients as possible. Cuttings use these micronutrients to knit mineral salts into vegetative root structures.

Although it is possible to mix your own solution, your safest bet is to buy a pre-mixed one from a garden center. Pre-mixed products are rich with the nutrients plants crave. Dilute them in warm water and use in place of water alone. Liquid and powdered seaweed is widely accepted as an organic fertilizer because it is full of trace minerals, micronutrients, and enzymes which are easily absorbed by plants. Some gardeners concoct their own fertilizer tea using guano, manure, and seaweed. Bat guano and worm castings can be added to soil or soilless mixes or soaked in water and fed as a liquid supplement.

Nitrogen, phosphorus, and potassium are also important to a cutting's success. They should be fed in a balanced proportion with nitrogen not exceeding 75 ppm. Once roots have formed, increase the amount of nitrogen and phosphorus you feed your cuttings.

Rule of Thumb: Feed rooted cuttings with 1/2 strength complete fertilizer.

Note: Parts per million (ppm) is another way of expressing a very small percentage. For example 0.001 percent = 10 ppm

Regardless of what nutrient solution you choose, it is important that you provide complete nutrition. For best results, feed your cuttings every two days. Most nutrient solutions will indicate how much to feed. In general, feed eight ounces of nutrient solution for every fifty cuttings. If you use a 100-cutting propagation tray, you will need sixteen ounces of nutrient solution.

The first feeding should occur when you are preparing your rooting medium. Pour the nutrient solution into the left side of the tray and gently move the tray back and forth so that the solution reaches all the medium. Slightly lift the left side of the tray so the solution travels downhill through the growing medium and to the right. Switch, lifting the right side of the tray up to move the solution to the left. Repeat the procedure lifting from the front and back. For the initial feeding, do not remove any excess solution. Wait fifteen minutes after feeding before planting your cuttings. Continue feeding in this manner once your cuttings are planted. Feeding this way may seem laborious, but it is an effective way to evenly disperse the solution without wetting the top of your soil or growing cubes. This is so important that large propagation facilities use shaker tables to water and feed their cuttings.

After the initial feeding, pour off any nutrient solution which remains standing in the tray. You want all the solution to be absorbed by the rooting medium. If your rooting medium absorbs all the solution, add more, continue the shaking procedure, and pour off any excess. Never leave nutrient solution standing in the base of your tray.

Once your cuttings have rooted and are ready to put down adventitious roots, their feeding requirements change. Now, instead of feeding your cuttings to encourage root growth, you want to feed nutrients which will encourage plant growth. About four weeks after taking your cuttings they should be ready to push out adventurous roots. Slowly let the rooting medium dry until adventitious roots have appeared. If the medium has dried out before these roots appear, apply a 150 ppm nitrogen solution with 75 ppm of phosphorous (1/2 strength solution). Apply this solution sparingly until roots show.

Once adventitious roots have been established, the cuttings are ready to be transplanted into larger pots and again, their nutrient needs change. For their first feeding in their new and expanded homes, apply a 300 ppm phosphorus solution with 100-150 ppm nitrogen. After that, change to a 300 ppm nitrogen solution for maximum vegetative growth. B^1 will also help ease the stress of the transplant and promote healthy growth.

Chapter Three
Rooting Mediums

Rooting Mediums

Rooting mediums serve several functions for plants; they support root systems and serve as holding-grounds for air, water, and nutrients. You can select between several mediums. Both soil and soilless mediums work well. In this chapter we will examine the advantages and disadvantages of each. Whatever medium you choose, be certain it possesses the following characteristics: it should be sterile, able to drain well, and able to retain air, water, and nutrients. These requirements rule out several options. Peat moss and compost are not recommended for rooting cuttings. It does not pay to try and skimp on rooting mediums. Even if you choose to grow in a soil mixture, you will need to purchase it or mix it yourself from sterile medium. Do not try to save a few dollars and take soil from your outdoor garden. This simply invites bugs and diseases already established outside to invade your indoor growing room.

Rockwool

Many gardeners prefer soilless rooting mediums to soil mixtures. In 1985, rockwool was introduced to the United States as an alternative to conventional soil and other soilless mediums. Rockwool is made of thin fibers of volcanic rock or a combination of volcanic rock, limestone, and coke. These threads of molten rock are compressed into layers and shaped for horticultural purposes into easy-to-handle cubes, sheets, or blocks of any size. Rockwool is also available in a granulated form. Rockwool cubes can be purchased in many garden stores (particularly those that sell hydroponic equipment) and are excellent for propagation because their vertically oriented fibers promote downward root growth. The cubes come attached in a sheet which fits neatly into a propagation tray. You can purchase pre-punched rockwool with holes for either seedling transplants or cuttings. Using this dry material eliminates much of the mess of gardening. There is no heavy soil to lug around or clean up. Be sure to buy horticultural grade rockwool. Although, rockwool is also used as insulation, the insulation grade is not recommended for gardening because the fibers run the wrong way and the rockwool has been treated with phenols to repel water and keep the insulation dry.

In addition to being easy to handle, rockwool is easy to control. It is sterile and consistent; there is no guessing game as there may be when you are preparing a soil mixture. Rockwool retains twenty percent air even when it is fully saturated. This helps provide a continuous oxygen supply to the roots. In addition, rockwool holds ten to fourteen times as much water as does an equivalent quantity of soil. It is easy to tell when your cuttings have rooted; you can often see the roots pop out the bottom of the cube. In addition, transplanting is

Cuttings growing in a flat of rockwool cubes.

effortless. Rockwool cubes are simply transplanted into larger pots. You do not need to worry about separating and potentially damaging the frail root system from its growing medium.

There are some drawbacks to working with rockwool. You need to remember to provide direct constant contact between the rockwool and the cutting. Pinch in the hole at the top of the cube to ensure that contact is made. When dry, rockwool can irritate your skin. Wear gloves and goggles if you must work with it dry or wet it to avoid skin-irritations. Rockwool also has a high pH level, approximately 7.8. Before using it, soak your rockwool overnight in a half strength fertilizer. Some growers who use Grodan rockwool report that they do not need to soak it overnight. Use an acidic fertilizer with a pH of 5.5 in order to maintain the pH of the rockwool at 6.5 or lower. Because rockwool stays so moist, algae see it as an inviting home. The algae does not consume a lot of nutrients, but it might attract fungus gnats. To avoid this, simply cover your rockwool with plastic to exclude light. To avoid stem rot, make sure that the plastic does not touch the stem. Or water cuttings from the bottom so that the top of the rockwool cube remains dry. The algae growth will not be as much of a problem for your cuttings before they root as it might be after. Before rooting, your cuttings will be partially shaded anyway.

Although rockwool is initially more expensive than soil (it can be up to ten times as expensive to root an equal amount of cuttings), it can be reused, making it quite economical. In between uses, remove any old root systems and sterilize the medium. Try to remove at least ninety to ninety-five percent of the roots to avoid potential clogs which prohibit effective draining. Sterilize the rockwool in a bleach solution after you have drained any standing nutrient solution. Mix one cup of bleach in five gallons of water and flood the rockwool for at least an hour. Then spend at least an hour flushing the sterilizing solution out of the medium (and any equipment you sterilized) with clean water.

Oasis Cubes

Like rockwool, Oasis cubes are easy-to-handle, pre-formed cubes which allow sufficient water, oxygen, and nutrients to meet the roots. Remember to pinch in the top of the cube around the cutting. Oasis cubes contain a small amount of 1-1-1 nutrient. This nutrient helps the cutting grow but the Oasis cubes still lend themselves to easy, complete control of the root environment. With Oasis cubes, it is easy to determine if your medium is moist enough. Simply place the bottom of the cube on the back of your hand. If it leaves a wet spot it is sufficiently moist. If not, it is time to water. Oasis cubes are

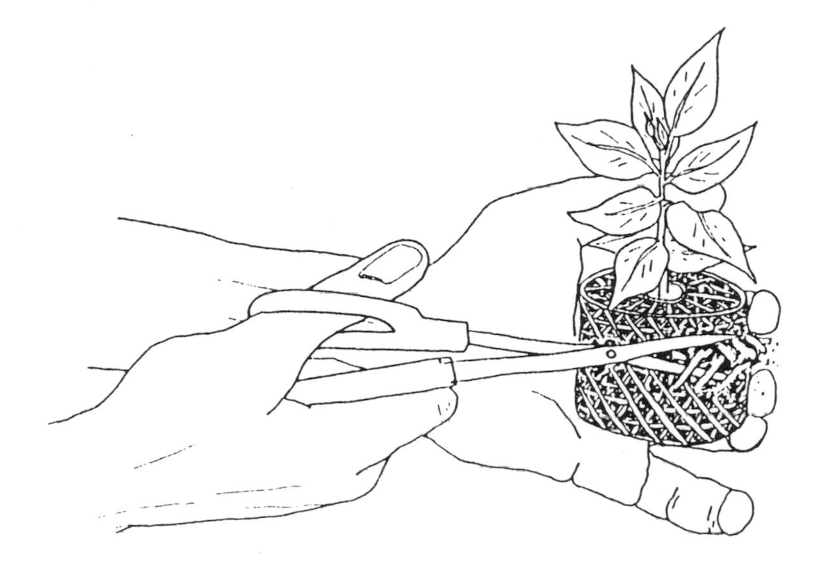

Trim the plastic netting from peat pellets before transplanting.

small so a hobby gardener tight on space can fit a lot of cuttings in minimal space without overcrowding. Like rockwool, the roots are easy to see. When roots have developed, simply place the cube in a larger pot.

Peat Pots

Peat pots work much like rockwool and Oasis cubes. They are small containers formed from compressed peat moss with expandable outside walls. You can purchase peat pots individually or in strips which are a series of small pots molded together for ease of handling. Jiffy-7's are made of compressed peat surrounded by a thin plastic net. When you purchase them, they will be flat pellets. When watered, however, they swell into little pots about two inches high and one and three quarters inches in diameter. Jiffy-7's are easy to store and are an excellent option if you want to start only a few cuttings. Jiffy-9's are similar in design to Jiffy-7's in that they are small pieces of compressed peat which expand when wet. However, Jiffy-9's are not covered with the plastic netting. Some gardeners find that because of this, they crumble easily when handled.

Before planting your cutting, make sure the peat pot is evenly moist. It is important to maintain constant contact between the cutting and the rooting medium. For both root cubes and peat pots, be sure to pinch the top in around the stem so that they are touching. If the top edge of the peat pot is left uncovered, it will compete with the cutting for water, wicking it away from the soil and evaporating it into the air. Peat pots, like rockwool, cubes, contain no nutrients of their own. This eases your job of controlling nutrient solutions. Finally, peat pots allow for effortless transplanting. When roots have formed, simply plant the peat pot in a larger pot. The peat pot will dry out and contract, exposing the roots, but causing very little transplant stress. To ensure an easy adjustment, slit the walls of the peat pot before repotting. This way if the roots can not force their way through the peat walls on their own, you will have provided an outlet. This is not necessary if you use Jiffy-7's whose open walls facilitate easy root penetration.

Soilless Mix: Perlite and Vermiculite

A mixture of perlite and vermiculite provides a great rooting medium. Perlite is a sterile white material of volcanic origin which has been expanded by heat, while vermiculite is mica which has been processed, expanded and popped like popcorn. Perlite is superior to vermiculite for aeration, but both retain water and nutrients well. Some gardeners soak the medium overnight in nutrient solution and then flush with fresh water to condition the mix. Perlite drains fast and does not promote salt build-up, an important quality for your delicate cuttings. When buying perlite look for medium or coarse, not fine grade. Vermiculite gives body to fast-draining soils and holds more water than perlite. For rooting, look for the fine grade vermiculite. If it is not available, crush the coarse or medium grades.

A mixture of two parts perlite to one part of vermiculite is an inexpensive and effective rooting medium. Be sure to wear gloves when making the mixture. Although you want to avoid breathing in the mixture, it is not recommended that you wet it. Mix outside, wear a respirator and lightly mist the dusty ingredients. Once prepared, distribute the mix into your propagation trays, making sure that each container is filled with at least one cubic inch. Many gardeners have had great success using this mixture. It retains water well, (although it can not compete with rockwool), is well aerated, is usually neutral on the pH scale, and is inexpensive. In addition, because vermiculite and perlite contain virtually no nutrients of their own you can easily control the root environment with nutrient and root hormones.

A perlite and vermiculite mixture has several disadvantages. Mixing it is time and labor intensive. In addition, because the mix is so lightweight, larger plants tend to fall over due to a lack of support. These plants must be staked or artificially supported.

An alternative to this rooting medium is a mixture of equal parts coarse sharp sand and vermiculite. The grains of sharp sand are rough to the touch, as opposed to soft sand which does not drain well and is too fine for our purposes. Sand you find on an ocean beach is not an acceptable growing medium because of its salt content. A sand/vermiculite mixture possess similar qualities to the vermiculite/perlite compound. Fine grade vermiculite can also be used on its own but many commercial seeding growers use a half sterilized peat moss half vermiculite mix. When reused, vermiculite breaks down in to a fine gooey clay, making it unacceptable for a medium.

Ferticubes apply the principle of peat pots to a perlite/vermiculite mixture. Ferticubes are made of sphagnum moss, perlite, vermiculite, and added nutrients which have been compressed into one inch square blocks.

Coconut Fiber

Coconut fiber is derived by immersing coconut husks in water for several weeks and then extracting the fiber mechanically. This product is catching on rapidly as a rooting medium for cuttings. It has a high moisture and air retention capacity and is packed with beneficial microbes. This product is available in briquettes and bales. The pH averages about 5, so the medium should be conditioned with a fertilizer solution before using as a growing medium. Coconut fiber can be mixed with perlite or vermiculite for added drainage or water holding capability.

Soil Mixtures

The most important thing to remember about rooting in soil mixtures is sterilization. Before a cutting has rooted, it has little ability to co-exist with any disease-causing organisms. Organic material provides an ideal home for the growth of living organisms which can hinder the rooting process for cloning. Avoid soil, unsterilized peat moss or compost. If you insist on using soil, be certain to use sterile, high quality potting soil. You can also sterilize soil in an oven or with steam, but this process can be time consuming, messy and can have an unpleasant odor.

There are numerous other mediums that also work to root cuttings. Expanded clay, Dutch corn and foam are just a few examples. Each medium has its strengths and weaknesses. The ideal medium is the one that works best for you!

Chapter Four
Rooting Hormones

Rooting Hormones

At the turn of the century, gardeners would split the cut end of their cuttings to place wheat seeds inside the cut. As the wheat seed absorbed water and germinated, it produced and released growth promoting substances. Cuttings rooted more easily and with more vigor. This practice became obsolete by 1940 due to the introduction and wide-spread acceptance of synthetic rooting aids known as rooting hormones or auxins.

Rooting hormones have become such a crucial element in propagation today that cuttings are categorized by how well their root formation responds to auxins. If auxins are unnecessary for rooting, cuttings are said to be easy to root. Difficult to root plants require a high concentration of rooting hormone. If a cutting fails to root, even after a rooting hormone has been applied, it is categorized as recalcitrant.

Although some gardeners forego rooting hormones, most choose the added security auxins provide even with easy to root softwood cuttings. Commercial rooting hormones are used to supplement a plant's natural hormones. When you think about the process of asexual propagation, it is fascinating to consider the process a cutting must undergo to root and become a healthy plant. Remember, in propagating plants we are asking a portion of a plant, be it a stem, root, leaf, or tip, to develop a root system strong enough to support new plant growth. In order for this to happen, a cutting must alter its production mode. Instead of producing stem cells, a cutting must begin to produce undifferentiated cells and then root cells. Rooting hormones promote undifferentiated growth and therefore stimulate and assist the rooting process. Auxins not only speed up the rooting process, but they also foster strong plant growth. Remember, fast rooting cuttings produce healthy, strong plants.

Although there are several which stray from the norm, most rooting hormones available on the market today are made from one of, or a combination of, three chemicals which are known to stimulate undifferentiated growth. 3-indolebutyric acid (IBA) is probably the most widely used because it is relatively safe (the dust is hazardous if inhaled) and effective for promoting root growth in many plants. Some commercially available products use only IBA while others use it in conjunction with either Alph-naphthaleneacetic acid (NAA) or napthaleneacetamide (NAd) which are closely related chemicals. 2,4-Dichlorophenoxyacetic acid (2,4-D) is also used in low concentrations as a root stimulant. More commonly it is used as an herbicide. Some manufactures add fungicides to their products to further promote rooting and to ward off damping-off disease.

Captan, benlate, thiram, and maneb are fungicides commonly added to rooting hormones.

Rooting hormones come in powder, liquid, or gel form. Each has its advantages and disadvantages, although many hobby gardeners and most production propagators choose liquid or gel over powder. Liquid and gel products allow you to apply the hormone in an even coat over the cutting, while the powder versions often adhere unevenly. This uneven distribution of hormones can negatively affect root growth. Gel and liquid rooting hormones also have the distinct advantage of helping to prevent embolisms.

If air becomes trapped in the stem an embolism can form and eventually cause your cutting to die. Gels work particularly well to seal cuttings immediately after they have been cut. This reduces not only embolisms, but also shock and infection. Because there is a risk of embolisms even when you use liquid or gel, be certain to dip your cutting quickly after making the cut to prevent air from getting trapped in the plant tissue. To completely eliminate the threat of an embolism, take the cutting, put it in a bowl of water and make the actual cut under water. If an embolism occurs, it may be difficult to detect. The plant may look healthy from the rooting medium up up for a month or two. All of a sudden the plant will fall over because the stem rotted at the base of the cube. Upon breaking the cube apart you are likely to find that the stem is rotten.

Most liquid hormones must be diluted by the gardener in a mixture of water and/or alcohol. You will find instructions on the packaging of your purchase. In general, to make a 1000 ppm concentration, add 1 gram (.035 ounce) of rooting compound in a one liter mixture of half water and half alcohol. For a 5,000 ppm solution, dissolve 5 grams (.175 ounce) of compound in 1 liter of water and alcohol mixture. The manufacturer will indicate different concentrations for different plants. Generally softwoods are easier to root than hardwoods and therefore require a lower concentration of hormone. Although gels and liquids are less hazardous to your health because they eliminate the possibility of inhaling toxic dust particles, some propagators are not comfortable mixing these chemicals at home. If you do not want to handle the chemicals, many pharmacists will mix them for you. When using liquid or gel compounds, pour an appropriate amount into a separate container so not to contaminate the bulk of your solution. Dip in this container and discard any remaining solution or gel.

There are two methods of applying liquid rooting products. Some gardeners soak their cuttings in a diluted solution for twenty-four hours. Generally these solutions contain between 20 and 200 ppm IBA or NAA. Results using this method vary because of differing environmental factors. During the twenty-four hour soaking period, environmental conditions such as temperature, light, and humidity affect the ability of a cutting to absorb the root promoter. More concentrated solutions of IBA and NAA, 500 to 20,000 ppm, can be used for the quick dip method. Because of the high concentrations of synthetic hormones, cuttings which are dipped for only five seconds can uniformly absorb the compound. To determine the rooting hormone concentration in ppm, multiply the percentage listed by the manufacturer by 10,000. For example, a product with .9% IBA contains 9,000 ppm IBA.

Synthetic rooting hormones sold in powder form are a mixture of talc and IBA and/or NAA. Powder rooting hormones tend to be less expensive than their liquid and gel counterparts. In addition, because they do not have to be mixed, they are less labor intensive. Simply roll the moistened end of your cutting in the powder. Do your best to apply a thick, even coat. Like liquid rooting hormones, powders can become contaminated. Do not dip directly into the original packaging. Pour a small amount into a separate container and throw away any excess. If you do choose powder hormones, you can increase your success rate and consistency by doing two things. Tap or scrape excess powder off the cutting because it can hinder rooting and growth. Secondly, make sure that you have created a large hole in your rooting medium. If the hole is too small you are likely to scrape off some of the powder as you insert the cutting.

Rooting hormones (liquid, powder, and gel) are marketed under many brand names. Although they range in price and exact formulation, for the most part they all rely on IBA, NAA, or a combination of both. If you need advice on selecting one appropriate for your needs, ask at your local garden supply center or write to the manufactures for more information.

Algimin is a liquid seaweed product manufactured by Maxicrop, USA (P.O. Box 964 Arlington Heights, Illinois 60006). Unlike most of its competitors, it does not use either IBA or NAA. Although it is not specified solely for propagation, many organic gardeners rely on it because they have found it to be an excellent natural root stimulator. This product is made from Norwegian kelp (a seaweed product) which provides a host of trace minerals and growth stimulants easily absorbed by plants. The manufacturer recommends soaking your cuttings overnight in a solution of 2 ounces Algimin to 1 gallon of water. After planting, continue watering with this solution.

Clonex (distributed through Superiors Growers Supply 4870 Dawn Avenue East Lansing, Michigan 48823) was the first cloning gel introduced to the market in the United States. In the gel base is a blend of 7 vitamins, 11 minerals, 2 anti-microbial agents, and 3,000 ppm rooting hormone. The combinations of these elements achieves several things. The cutting is fed and protected by vitamins and minerals; hormones stimulate root formation; and anti-microbials ward off fungal diseases. Because it is a gel, it seals the tissues of your cutting upon contact, thereby reducing the chance of infection or embolisms. Use Clonex Purple for softwood and Clonex Red for hardwood cuttings.

Dip N' Grow (Astoria-Pacific, Inc. P.O. Box 830 Clackamas, OR 97015) contains IBA, NAA, and an anti-bacterial and anti-fungal agent. The manufacturer of this liquid rooting concentrate says that it has increased cutting survival rate by up to 60% for propagators nationwide. Since you can mix any concentration by varying the dilution of Dip N' Grow and water, it is effective for both softwoods and hardwoods. Dip N' Grow tends to be less expensive than its competitors, averaging a penny per 100 cuttings.

Earth Juice Catalyst (OGM, POB 3442, Chico, CA 95927) is an organic product derived from oat bran, kelp, molasses, vitamin B complexes, amino acids, hormones and low levels of nutrients.

Frank's Root grow (FMCI Hydroponics, 480 Guelph Line, Burlington, Ontario, Canada, L7R-3M1) contains Indole butyric acid, Naphthalene acetamide and naphthalene acetic acid as active ingredients. These ingredients are vibrated into lanolin so they do not wash off the plant.

Hormex (Brooker Chemical Corp. P.O. Box 9335 North Hollywood, CA 91605) is an IBA based powder which is available in six different strengths ranging from 1000 ppm to 45,000 ppm. The latter is intended for extremely difficult to propagate plants and will severely

Pour the (diluted) liquid rooting hormone into another container before using it to guard against contaminating the stock solution.

burn most cuttings. Read the directions on the labels and choose carefully. Hormex also produces a product which blends IBA, NAA, and Vitamin B1. It can be diluted and used for watering.

Hormodin (Merck & Co., Inc. P.O. Box 2000 Rahway, NJ 07065) is a powder which comes in three strengths, 1000, 3000, and 8000 ppm. Its main active ingredient is IBA.

Nitrozyme (Pure Food Hydroponics, 3385 El Camino Real, Santa Clara, CA 95051) is an extract from a seaweed-like plant (*Ascophylum Nodosum*) and contains numerous hormones including cytokinins, auxins, enzymes, giberellins and ethylenes. The manufacturer recommends spraying the product on mother plants two weeks before taking cuttings.

NutriRoot (Western Water Farms, 103 - 20120 - 64th Avenue, Langley, BC, CANADA V2Y-1M8) This high quality rooting gel can be used for micropropagation, tissue culture, stem and leaf cuttings for hardwood and softwood stock. Consists of a balanced blend of trace elements, sugars, vitimins, minerals and growth regulating hormones.

Olivia's Cloning Solution (Olivia's Solutions, Inc. P.O. Box 887 Calistoga, CA 94515) makes a very popular cloning solution available in liquid and gel form. Gardeners report a 90-100% success rate using Olivia's cloning product (theystimulate root growth and feed the cutting simultaneously). The product works well for softwood, semi-hardwood, and some hardwood plants. The manufacturer recommends watering with the liquid solution once cuttings are planted.

Powerthrive (B & B Hydroponics, 375 MacArthur Avenue, Ottowa, Ontario, Canada K1L-6N5) is a natural root stimulating product containing kelp, Vitamin B^1, cytokinin, and many other growth stimulants, hormones and nutrients.

Rhizopon AA (distributed by Hortus USA, 245 West 24th Street, New York, NY 10011) is imported from a Holland-based manufacturer, Rhizopon B.V., the world's largest company devoted to researching and manufacturing plant rooting products. This product uses IBA as a root stimulator and is available in three different strengths in both powder and water soluble tablets. The tablets can be mixed in water and/or alcohol to form concentrations ranging from 500 to 20,000 ppm.

Rapid Root Liquid (Higher Yield, 29211 NE Wylie Rd., Camas, WA 98607) has a very high concentration of IBA and NAA. - It is the strongest liquid concentration available!

Several popular rooting hormones.

Rootex-L (Irelands Hydroponics, 31 Main St., Kinglake, 3763, Australia) This general purpose liquid rooting hormone has been widely used in nurseries for years. The manufacturers sight convenience, versatility and a high strike rate as the main features to this IBA-based product.

Rootone (The Chas Lilly Co. Portland, Oregon 97283) is a mild powder formulation which uses 2000 ppm NAd (a close relative to NAA) as its main root promoter. Thiram is added to the product to protect against fungal disease.

StimRoot - (Tom Thumb Hydroponics, 4055 Harvester Road, Unit 15, Burlington, Ontario, Canada l&l-5Z7) is a root inducing liquid hormone containing both IBA and NAA.

Superthrive (Vitamin Institute P.O. Box 230 North Hollywood, CA 910603) is not designed specifically for cuttings, although many gardeners have found it to be a good root stimulator and presoak prior to taking cuttings. Although its exact formulation has not been revealed to the public, it is known to contain NAA and a substance similar to B^1, as well as 50 other vitamins and hormones. The combination of ingredients helps induce roots, minimize transplant shock, and encourage healthy and strong plant growth. To use, simply mix a few drops per gallon of water. You can use it as a rooting hormone and for subsequent watering. The same company produces a similar product especially for cuttings called "Cutstart".

Wilder's Rooting Glue (Wilder Agriculture 4188 Bethel Wilmington Road New Wilmington, PA 16142) is a semi-solid solution, similar in consistency to a gel. Because of its sticky consistency, Wilder's Rooting Glue sticks well to the end of your cuttings, thereby making hormones immediately available to plants and curtailing damping-off disease. This rooting compound combines IBA and NAA in a base of several anti-fungals, including sulphur and streptamyacine. Wilder's can be used on either soft or woody cuttings. In addition to their rooting compound, Wilder manufactures a plant food known as CRC (Clone Rooting Concentrate). The plant food contains the same ingredients as the Glue, but in diluted concentrations. Wilder's recommends using CRC for the first two to three weeks of a plant's life to ensure vigorous rooting. If used in combination, Wilder's predicts a 95-99% rooting success rate with their products.

Wilson's Roots (Tom Thumb Hydroponics, 4055 Harvester Road, Unit 15, Burlington, Ontario, Canada l&l-5Z7) is a rooting hormone containing IBA, NNA, 5-ethoxy-3-trichlormethyl-1,2,4-thiadiazole and a fungicide. It is available in both powder and gel form. The supplier recommends the gel.

Wood's Ready to Use Rooting Compound is a liquid rooting solution manufactured by American Agriculture (9220 SE Stark Portland, OR 97216). Wood's relies on both IBA and NAA to stimulate root growth. After 30 years of testing, Wood's has developed a product which works particularly well with difficult-to-root cuttings such as trees, rhododendrons, lilacs, and other woody plants. Because Wood's solution is easy to synthesize, it is immediately available to plants. Follow the directions on the label to mix and use Wood's rooting compound. Wood's also markets an anti-wilting spray which can be used in conjunction with their rooting compound. The anti-rooting compound should be sprayed on leaves to prevent plant dehydration.

Chapter Five
Taking Cuttings

Equipment

Although cuttings can be grown outdoors in shady areas, our focus is on growing cuttings indoors. To get started, you will need some very basic and thankfully inexpensive equipment, much of which you can make yourself. A propagation tray, purchased from any garden supply store for a couple of dollars, works well for cuttings. Some gardeners start their cuttings directly in pots to avoid transplanting them later. If you choose this approach, plastic pots are preferable to clay pots as they do not absorb moisture and compete with the plants. You will need a cutting implement. A knife, X-acto knife, single edged razor blade or sharp scissors all work well. There are two important specifications for your cutting implement; it should be sterilized before use and in between cuttings, and the sharper the better. A dull knife can tear the tissue on the plant. Remember, taking cuttings is the most traumatic experience a plant will undergo in its life. You need to do all you can to alleviate plant stress.

Successful gardeners dip their cuttings in a rooting hormone before planting them. Refer to the previous chapter for an in-depth discussion about rooting hormones. Once your cuttings are planted securely in their rooting medium, you need to ensure ideal growing conditions. Cuttings thrive in moist (not soggy), warm environments. Depending on the environment where you intend to grow, you might need to buy some equipment to help maintain proper heat and humidity levels. You can choose from both home-made or store-bought alternatives to create a warm and moist propagation area. In addition, and again depending on the pre-existing environment, you might want to consider purchasing a lighting unit. In total, with some creative thinking, it is possible to establish a propagation area for under (US) $100. In chapter one you will find a complete discussion about creating and maintaining an optimum growing environment.

If you have a garden room in your house, that is ideal. If you do not, it is easy to establish a space to root your cuttings. Choose a spot which is light, but which does not receive direct sun, as direct sun can dry out your cuttings. The area should be clean, and out of the way of heavy traffic. The greater your ability to control your cutting's environment, the more successful you are likely to be.

The final pieces of equipment you will need are perhaps the most essential - a notebook and writing utensil. It is important to keep data on the cuttings you take. Record when you took them, from which mother plant, what rooting hormone, if any, you used and the growing conditions you provided for the plant. If you

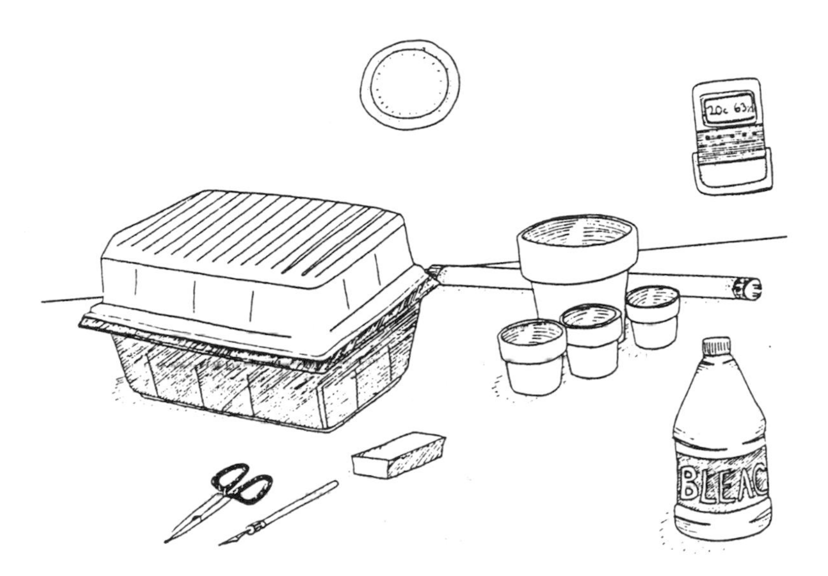

Lay out all of your supplies before starting to take cuttings.

keep accurate notes you will be able to identify with which plant varieties you are successful, and which varieties require more attention. You will also be able to record the different factors that effect the cuttings' success – lighting, humidity, temperature, nutrient solutions, rooting medium and rooting hormones–and adjust your growing environment accordingly.

Choosing a Mother Plant:

The mother plant, also referred to as the stock plant, is simply the plant that produces the cuttings. Cuttings replicate the mother plant in every way: size, color, taste, smell, strength. Environmental factors will affect the cuttings. You may experience some individual

The ideal mother plant.

characteristic variation. Some cuttings may root twice as quickly as others taken from the same plant. This is almost always the result of environmental factors. Are the cuttings overcrowded? Are some receiving more light than others? Did you inadvertently forget rooting hormone for one, or apply less to another? Is the humidity level consistent for all your cuttings and are you watering equally? All of these seemingly minor details can affect the rate at which your cuttings grow. For the most part, however, a cutting will follow the same path as its parent plant and will develop into an identical cutting. *It is therefore imperative that you choose a good mother plant.*

Choose a mother plant that is genetically mature. For most plants this means it must be at least two months old. Your mother plant should ideally possess all of the following qualities:

1. It should have short, compact, bushy growth. Taller plants tend to block out too much light, shading the ones beneath. Limited light and overcrowding only add stress to this already traumatic time in a plant's growth.

2. The mother plant should be disease-resistant. A diseased plant will produce diseased cuttings. It is easy to be tempted by the last healthy growth on an otherwise unhealthy plant, especially if it is one of your favorites. Taking a cutting from such a plant will only bring you disappointment. The cutting will be sickly and you will now own two failing versions of your favorite plant.

3. Your mother plant should have a history of heavy flower and fruit production. Taking a cutting while it is flowering or bearing fruit will slow the plant's growth. The cutting would have to reverse the flowering cycle and revert to vegetative growth. Look for a mother plant which flowers heavily, make a mental or written note, and take your cutting from this plant during its vegetative stage. If you do take a cutting from a flowering plant, it will root quickly but can then take up to a month to revert back to vegetative growth. This delay can cause additional stress to the cutting.

When taking a cutting from a mother plant always think of the mother's future growth first and foremost, unless you intend to dispose of the mother after the cuttings have been taken.

When you are preparing to take a cutting you must pre-treat the mother plant. Because a low nitrogen content in the stem speeds rooting, you should put your mother plant on a low to medium nitrogen diet. Many gardeners leach the soil where their mother plant grows with large quantities of water for several days prior to taking a cutting. Although leaching the soil with large quantities of water is an effective way to wash away nitrogen, you run the risk of creating very soggy soil. To avoid this risk, some gardeners leach the leaves as opposed to the soil. To do this you implement a practice referred to as reverse foliar feeding. Foliar feeding is a method by which you feed your plants via their leaves as opposed to their roots. Reverse foliar feeding is exactly the opposite. As opposed to feeding the plant via its leaves, you are cleansing it. Simply mist plants thoroughly every morning for a week with warm water. This process quickly washes the soluble nitrogen out of the foliage.

There is a trade off for having a low nitrogen content in you mother plant. For softwood cuttings, carbohydrates help to give stiffness to the stem, a desirable attribute which guards against stems bending when you are planting your cuttings. For this reason, some gardeners recommend maintaining a minimum of 150 parts per million (ppm) of nitrogen for the mother plant to help promote positive growth. Others argue that because the mother plant's growth is slowed as the nitrogen is used, carbohydrates accumulate. A high carbohydrate content enhances stiffness in branches and hastens root formation. Perhaps the best solution is to maintain a low nitrogen diet until a week before you intend to take your cuttings. At that time, begin leaching so that carbohydrates will accumulate and add rigidity to your branches.

To check the carbohydrate content of a cutting, dip it in iodine. The stronger the color absorbed by the cutting, the higher the starch (carbohydrate) content. This helps determine which area of the mother is best to cut.

You might also want to supplement your mother plant with phosphorus and potassium. A phosphorous diet of 75 to 150 ppm will restrict root growth in proportion to vertical height. This will help produce a full, bushy plant. If your mother plant is young and has not yet reached an ideal size, you can increase root growth by maintaining a 3 to 1 ratio between phosphorous and nitrogen. Add 300 ppm phosphorous and 100 ppm nitrogen.

Potassium helps to sustain growth and should be applied in equal parts per million as nitrogen.

A strong mother plant can withstand a lot. You can take many cuttings from the same mother plant at one time. Be certain not to remove more than sixty percent of the mother plant's vegetation. The plant, if properly cared for, will rebound with more vigor and bushiness within four to six weeks. For best results, retire the mother plant after four cutting sessions with four to six weeks between each cutting session. After this, the plant can become stressed and will therefore produce stressed cuttings. With each round of cuttings you take, select new mother plants. This way, you will always have healthy mother plants on hand. Your cuttings will be ready to give cuttings in two months, or as soon as they reach full vegetative stage.

Taking Stem Cuttings: Step-by-Step

Most hobby gardeners concentrate their propagation efforts on softwood cuttings. Softwoods tend to root more easily, therefore require less equipment, money, and time. Softwood refers to the soft, green, stem tissue in

Reverse foliar feeding.

plants. Most houseplant annuals are softwoods, as are the first season's growth of favorite outdoor perennials such as geraniums and chrysanthemums. Once the stems of such perennials mature into woody tissue they are referred to as hardwoods.

Having covered the basic issues of propagating (equipment, lighting, humidity, temperature, rooting mediums, and rooting hormones), we will now turn to the step-by-step details which will guide you through the process.

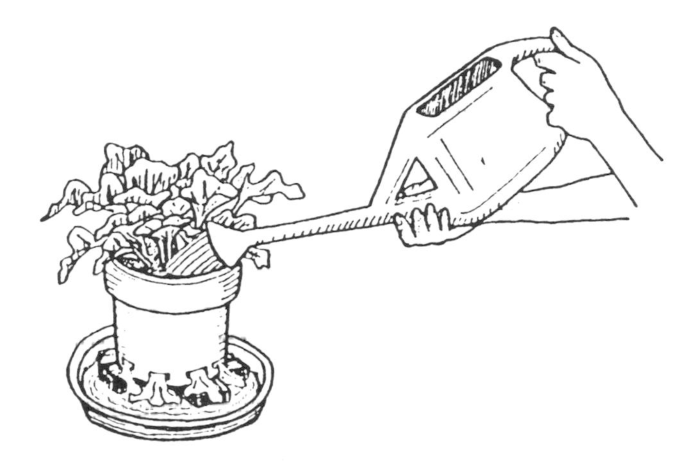

Preparing to take cuttings by leaching the mother plant

Stem Cuttings

1. **Preparation** - Some preparation is necessary before you can actually take cuttings. The following things should be done. Choose a mother plant which is at least two months old and 24 inches tall. (Some hydroponic gardeners take cuttings from month-old plants). Begin leaching seven days prior to taking cuttings. Make sure you have established a good growing environment. Check your equipment to make sure that everything is in working order. Your cuttings will need access to good light, heat, and water. Prepare your rooting hormone and rooting medium. Some rooting cubes must be rinsed in hot water before use to rid them of excess salt. Drain excess water from your cubes and fill your sterile propagation tray with rooting medium. If your medium is not pre-punched, create a hole in each block using a pencil or

Sterilize the clippers by dipping them in a jar of rubbing alcohol or bleach.

chopstick. The hole should be big enough so it won't scrape off hormone when you are inserting your cutting. The depth of the hole is dependent on the medium. To allow for healthy root growth, leave undisturbed at least half an inch of rooting medium at the bottom of the container. Once you take your cutting, timing is imperative. If you are using powder rooting hormones, you might want to have a container of water on hand. Storing your cuttings in water, or wrapping them with a moist paper towel will help prevent embolisms.

2. **Sterilization-** Sterilize all your equipment including your cutting implement, propagation tray, and humidity box with a mild bleach solution (1/2 cup bleach per gallon of water), making sure that you clean all the ridges and crevices of your propagation tray. Rinse every-

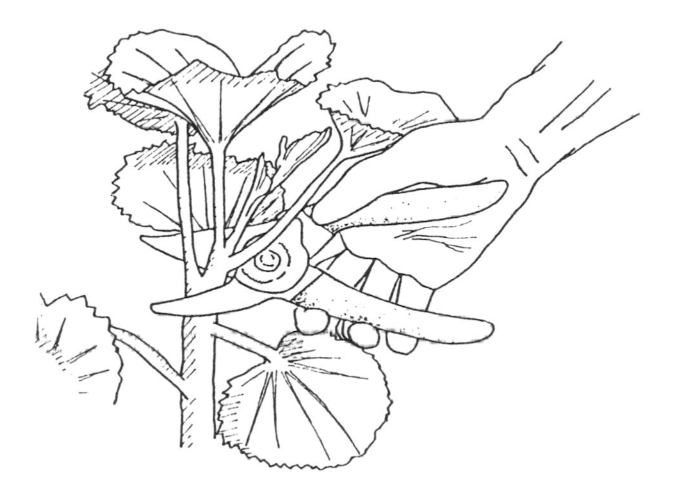

Cut the stem of the cutting on a 45 degree angle.

thing thoroughly, allowing clean water to flush your equipment for several minutes until no more chlorine smell is present. Air-dry the equipment. You can also use a lighter to heat and sterilize the edge of your cutting implement. Nursery-people around the country report that lack of sterilization is one of the most common causes for low success rates when propagating. Remember, your cuttings are stressed and in need of tender care. Germs and disease carried on your equipment can cause the cuttings to die. Do not smoke while taking cuttings. Wash your hands and equipment. Re-sterilize your cutting implement between every 10-15 cuttings.

3. **Timing-** Although cuttings can be taken any time of year, some gardeners report better results when they take their cuttings in the early spring (mid-March)

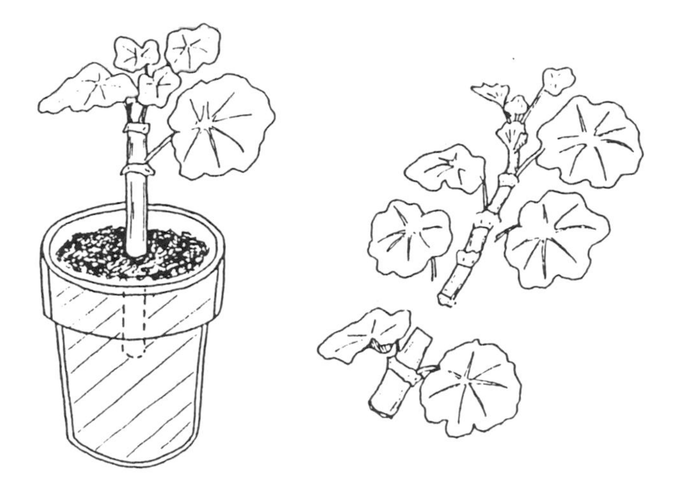

Remove leaves on the plant to get the stem ready for the rooting medium.

when plants are responding to increased daylight and actively growing. Some gardeners only take cuttings during a full moon. If you are growing under HID lights, you can take the cuttings any time of year. Take cuttings from the current year's growth. It is possible to take a cutting from a flowering plant, but for best results, choose a plant in its vegetative growth stage. If you want to cut from a flowering plant, remove all of the flowers from the branch that is being used for cutting stock. If possible, take your cutting in the early morning.

4. **Cutting-** You can use sharp pruning shears, scissors, an X-acto knife, or a grafting knife. *Make sure your cutting implement is sharp enough so it will not damage the end of the cutting or the mother plant.* Take your cutting from young growth which will be better able to develop

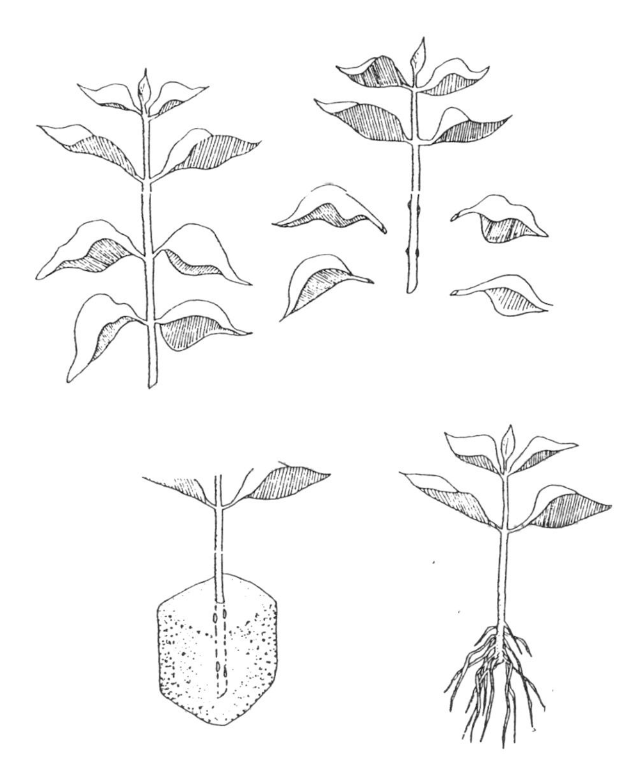

This drawing shows a cutting that has had the leaves removed, one stuck in rockwool and a final rooted clone.

roots than their older counterparts. Your cutting should bend without snapping. Cut your stem four to six inches long and between an eighth and a quarter inch in diameter. It is important that your cutting be the proper width. Excessively thin cuttings take a long time to root and stress the plant in the process. Cut at a 45 degree angle (some people recommend 60° to increase surface area), half way between leaf joints, referred to as nodes, making sure that your cutting has three to four sets of well-developed leaves and a sprout at top. Cutting between nodes,

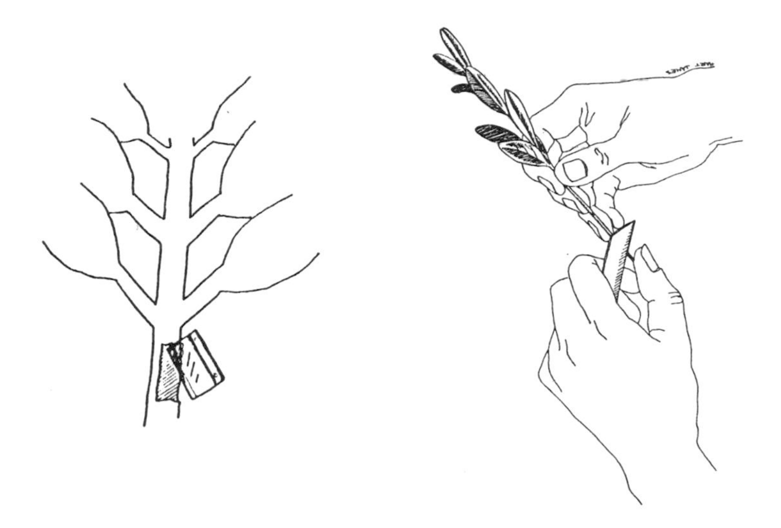

You can scrape the outer layer of tissue off the stem with a razor blade to speed root growth.

as opposed to directly above one as some gardeners recommend, assures that you will not damage the nodes. Cuttings taken from the top growing shoots of a plant will root well, provided they are thick enough. To decrease transpiration, you can also cut off 1/2 of each leaf. Continue taking cuttings from the outer growths of the mother plant. Although cuttings from the inner part of the plant may not have sprouts, they will root. Remember, do not remove more than sixty percent of the parent plant in one cutting session.

5. **Trimming-** Without gouging the stem, trim off the two sets of leaves at the bottom of the stem. This leaves the nodes, which contain the highest concentration of potential rooting cells, exposed. These exposed nodes will be planted so that they are covered by your rooting

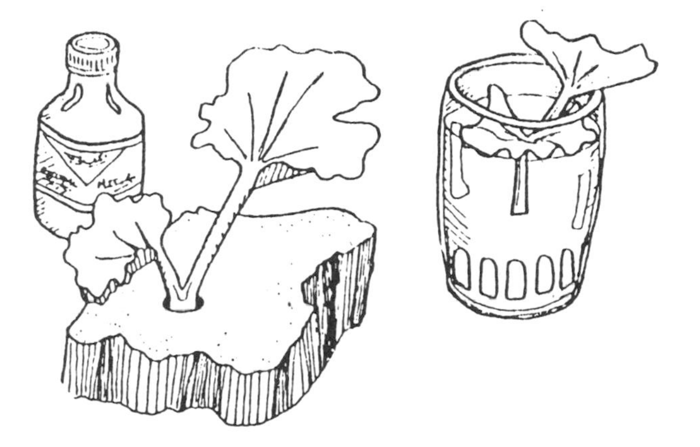

Dipping cutting in rooting hormone.

medium. Although two sets work best, you can rely on a single node (not recommended) if you have to. If the cutting has more than one large sun leaf, cut off all but one because if is difficult for the cutting to keep large leaves alive. Although most gardeners do this step after having cut the cutting from the mother plant, some recommend trimming before cutting. This allows you to transfer the cutting to the rooting hormone more quickly. Remember, timing is critical. You do not want air bubbles to enter your cutting. Place the trimmed stems in a bucket of clean, warm water or wrap them in damp paper towel so they do not dry out before you proceed to the next step. Some gardeners recommend leaving the cuttings in water overnight in low or no light. Cuttings should remain vertical because this allows for root-production chemicals within the plant to translocate to the base of the cutting.

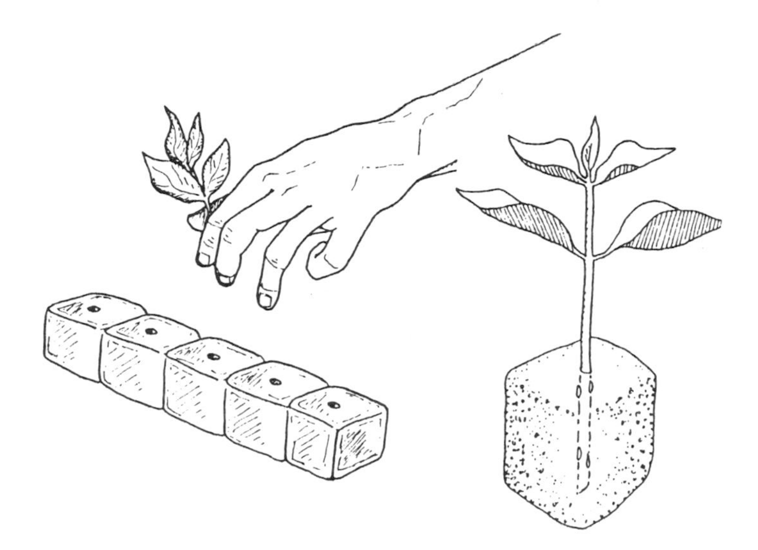

Planting cuttings in rockwool cubes.

6. **Wounding-** Some gardeners have adopted the practice of wounding (usually used for grafting) for stem cuttings. Although there are no definitive answers, there is much speculation about the benefits of wounding. The one thing we do know is that wounding takes time and is not practical when taking lots of cuttings. However wounding is believed to hasten root formation by exposing tissues with root-forming potential. In addition, wounded areas show a higher concentration of both natural rooting hormones and carbohydrates, both of which aid in the quality and quantity of roots produced. Wounded cuttings might also be able to absorb more water and rooting hormone from the rooting medium. To wound a cutting, simply make a knife cut at an upwards diagonal angle on the bottom surface of the cutting. The cut should penetrate half way through the stem thickness.

Rooted cutting in rockwool cube. Note the root growth penetrating the rockwool cube.

If necessary, use a match stick to hold the cut open. For thick-stemmed succulent plants such as geraniums, allow a callous to form before continuing. Place the cutting on newspaper and let it sit indoors for several days. Once a callous has formed, skip step 7 and proceed. Some people also split the stem vertically for an inch or so to give more surface area increase water and nutrient uptake.

7. **Scraping-** Using a razor blade, gently scrape a layer of skin off the portion of the stem that will be covered by the rooting medium. Scraping the stem disrupts the cells on its surface and hastens their change to rooting cells. This step is optional and not always necessary.

8. **Rooting Hormones-** Depending on the type of rooting hormone you are using, the process will vary slightly at this point. Roll the end of the stem in rooting

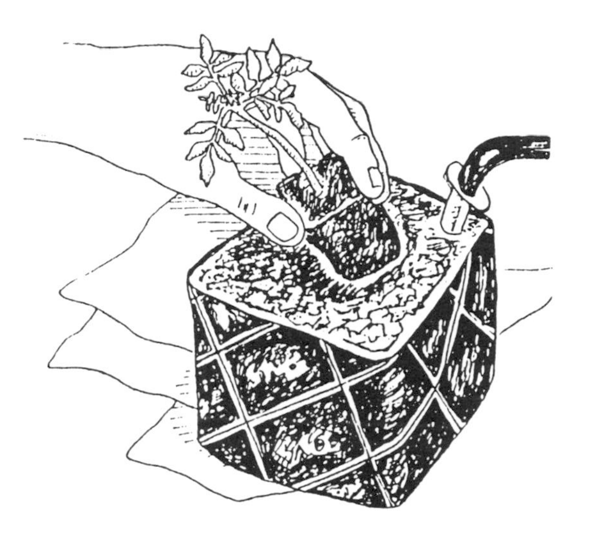

Transplanting a rockwool cube into a larger rockwool cube.

powder; dip in the liquid or gel varieties. Transfer the cutting quickly from the cloning solution to the rooting medium to avoid embolisms. If you are using gel, transfer time is less important because the gel seals the stems upon contact. If you are soaking your cutting overnight in liquid hormone, do so and continue from step nine tomorrow.

9. **Planting-** Be gentle when you are inserting your stem so as not to disrupt the rooting hormone. Place the stem securely in the rooting medium. If you are using cubes, do not allow the stem to poke through the bottom. Gently pack the rooting medium around the stem. Water your cuttings with a diluted nutrient solution, or a mild B^1 solution until the medium is evenly moist, but not soggy.

Tugging on plants breaks roots!

10. **Aftercare-** Follow the guidelines for environment control discussed in Chapter One. Remember, proper lighting, humidity, and temperature are essential for healthy cuttings. Remove rotten leaves from your cuttings. In addition throw away any cuttings which look diseased. Some cuttings may wilt for a few days. If cuttings are still wilted after a week, they are unlikely to live to be strong plants and should be discarded.

11. **Transplanting-** It will take several weeks (one to four) for a cutting to develop a strong root system. If you are using peat pots or cubes, you will be able to see the roots developing. If you can not see the roots, pull gently on the base of the stem after two weeks. If there is resistance, there are roots. The tips of leaves may also

turn yellow once roots start growing. Once a cutting has a strong root system, it must be transplanted. Next to taking a cutting, transplanting is the most stressful experience a plant lives through. Because the young roots of your cutting are delicate, you need to apply the same care you did while taking and raising the cutting.

Water the cutting to be transplanted with half-strength vitamin B^1 one or two days before transplanting. Transplant late in the day to provide the plant recovery time during the night. If you are using cubes or peat pots, simply transplant the cutting into a larger container and fill in around the pot or cube with soil. If you rooted in a perlite/vermiculite mixture, prepare the medium in a larger container, creating a hole into which the root ball will fit. To avoid unnecessary stress and adjustment, transplant into the same type of soil you used as a rooting medium. If you have rooted cuttings in soil and pots, saturate with water then roll the pots between your hands to break the sand away from the pot sides. Place your hand over the top of the pot, with the stem between your fingers. Turn the pots upside down and allow the root ball to slide into your hand. You may have to tap the pot lightly to free up lower roots. Place the cutting in the hole in the larger container, making sure that all roots are growing down, and pack the soil firmly around the root ball. Roots need to be in direct contact with the soil in order to absorb nutrients and water.

To ease plant stress, water with half-strength vitamin B^1 until the soil is saturated, but not soggy. Water helps transplants by packing in the soil so it remains in contact with the roots. Transplants should be fed low levels of nitrogen and potassium and large amounts of phosphorus. Many gardeners use a "bloom" fertilizer mix. If you use rich potting soil, it will provide enough nutrients for a month. Clones in soilless mixtures require fertiliza-

tion within a day or two of transplanting. Transplants prefer low filtered light. Fluorescents are preferred to HID lights. If you use HID lights, be sure to provide partial shade for the transplants or raise the HID lamp to 3 - 5 feet above cuttings.

Our focus thus far has been on softwood stem cuttings. In this section, you will discover many methods of asexual propagation. Although many gardeners agree that softwood stem cuttings are the easiest to propagate, none of the methods discussed in this chapter are difficult. The beginning of this chapter provided you with a detailed outline you will need to refer to as you experiment with the alternative rooting methods described below.

Semi-hardwood and Hardwood Stem Cuttings

Semi-hardwood and hardwood stem cuttings are taken using virtually the same process outlined above. You must pay attention to several variations in procedure because hardwood and semi-hardwood cuttings are older than softwood cuttings.

Semi-hardwood cuttings can be taken from fuchsias, camellias, and gardenias. Like softwood cuttings, semi-hardwood cuttings should be taken from the current season's growth. The main difference between softwood and semi-hardwood stem cuttings is timing. Take semi-hardwood cuttings in the fall. Leave a sliver of bark or stem known as a "heel" on semi-hardwood cuttings. Follow the instructions for propagating softwood stems.

Hardwood stems are easy to identify because they are last season's growth which has hardened into wood. This method is used primarily for outdoor flowering trees and shrubs, although gardeners in the South who contend with stem and root rot will find that it is a helpful alternative to propagating houseplants. Citrus fruits and poinsettias root particularly well using this method.

Stems can be longer for hardwood cuttings than for softwoods. Choose a section that is 5-10 inches long and cut it at an angle slightly below a node. Cut the top of the stem off one half inch above the top node. Make this cut straight so you can easily identify the top from the bottom. Use a rooting hormone at a strength recommended for hardwood cuttings. Plant the stem with at least 2 inches buried in the rooting medium. Bottom heat will speed the rooting process. Follow aftercare and transplanting instructions for softwood cuttings.

Leaf Cuttings

Leaf cuttings work particularly well for African violets, rex Begonia varieties and Peperomias, but you can experiment with leaf cuttings from most succulent, fleshy-leafed plants and cactus. Different plant varieties require slightly different methods.

1. **Preparation-** Prepare the growing environment as you would for a softwood stem cutting. Give the mother plant a large drink of water several hours before you intend to remove the leaf.

2. **Cutting-** Choose a large, healthy, mature leaf. Use a sharp knife, razor, or scissors and cut the leaf off at the base of the stalk leaving an inch of stem. For African violet leaves, be sure to leave at least two inches of stem attached to the leaf. African violet leaves should be kept whole. For other varieties, cut the leaf into wedges taking care to leave a vein in the center of each leaf cutting. Each cutting should also have a piece of the sinus (where the leaf and stalk meet) at its tip. Leaves can be divided into several plantable sections. For instance, a rex begonia leaf will yield up to five v-shaped segments.

3. **Rooting Hormone-** Dip the sinus tip into your rooting hormone.

4. **Planting-** Plant the leaf or leaf segment from a 30° angle to vertical so that one-third of it is covered by your rooting medium. Make sure that the sinus-dipped end is down and gently firm the rooting medium around the leaf. The process will not work if you plant the leaf upside down. If you do not use rooting hormones, cut the bottom at an angle so you can identify it. Leaves

Leaf cutting.

should be planted at least an inch apart to allow for growth. Segments can also be placed flat on the growing medium as long as they are weighted down. Put the sinus end under the growing medium and gently cover. Water the leaves lightly.

5. **Aftercare**- Aftercare for leaf cuttings is similar to that for stem cuttings, it just lasts longer. Provide a warm, moist environment. Make sure not to dislodge the fragile leaves or leaf segments when watering. Roots will develop from the leaf wedge in several weeks, but it may take months for new leaves to appear.

6. **Transplant**- When the new leaves are about the size of your thumbnail, move the plantlets from trays to small, two and one-half inch pots. Transplant as you normally would, making sure to divide any clumps which contain several plantlets. Each platelet you transplant should have a healthy set of roots. Remove the original leaf or fragment if it has not already rotted. When re-planting, be careful not to cover the crown and

growing tip of the tiny plant. Keep the plant in shaded light for several days. When the roots are too large for their containers, transplant again into larger containers.

Alternatively, a leaf cutting can be taken by cutting the veins at the junctions of the main vein and laying the whole leaf flat in rooting medium. The bottom of the leaf must be down and in contact with the rooting medium. Use stones or bobby pins to hold the leaf down and plant the cut stem firmly in the medium. Follow aftercare instructions described above. In several weeks, plantlets will form on top of the leaf at the cut veins.

Root Cuttings

Plants with thick, fleshy roots or ones that produce suckering shoots are good candidates for root cuttings.

1. **Preparation**- Prepare a growing area which will provide bottom heat, humidity, and light. You can use either trays or individual pots for root cuttings.

2. **Cutting**- Simply cut a portion of the root from the parent plant. Cut this portion into smaller sections, ideally 1-4 inches long. Root cuttings work best if they are 1/4-1/2 inch in diameter. Cuttings taken in the fall or early winter will benefit from the stored food in their roots.

3. **Planting**- You can either lay the roots flat in a tray, or plant them vertically in pots or a tray. Many gardeners prefer to plant the roots vertically and report better plant formation from this method. If you plant them flat, cover them lightly with your growing medium. The top covering should be 1/2 to 1 inch thick. Clearly cubes will not work for this method. If you want to use rockwool, look for the granules.

4. **Aftercare and Transplanting**- Follow aftercare and transplant instructions for softwood stem cuttings.

Division

Propagating plants through division is an easy way to increase your plant stock. Many houseplants including ferns, African violets, clivias, and lily-of the-Nile have more than one crown and can therefore be divided. Divide foliage plants in the early spring before new growth begins. Flowering plants should be divided during their dormant period.

To divide your plant, roll the plant pot between your hands to loosen the soil. Turn the pot upside-down, placing a hand over the top of the pot with the plant stems between your fingers. Tap the pot to free the plant and soil if it is root-bound. Shake off excess soil. Separate the root ball and parent plant into as many sections as there are stems or crowns. This can usually be done with your hands, but pot bound plants which are difficult to separate might require the help of a knife. With difficult to separate plants, rinse the root ball in water to remove the soil from the roots. Try to keep as many roots as possible attached to their original stems. Plant each division in a separate container. Planting should be quick so not to dry out the roots. Water generously and frequently. Keep the plants in diffused light until they show new growth.

Although propagating by division is easy and requires little equipment, it is not an ideal method if you are looking for large yields. Most plants can only be divided into two to five separate crowns. This is not many copies of the plant when compared to stem cuttings where many parent plants will produce 20 - 25 cuttings per cutting session.

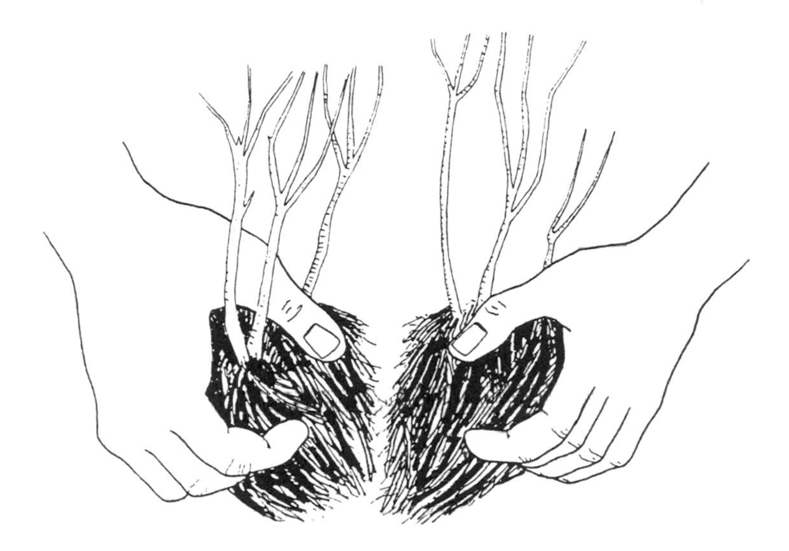

To divide roots, simply pull roots apart.

Layering

Simple layering is an easy way to propagate small numbers of plants. Because it is labor and space intensive, this method is probably not a good choice if you want to produce many cuttings. If you want to increase only a few plants and are having difficulty rooting them from cuttings, layering is a good alternative. Layering involves rooting a shoot which is still attached to the mother plant. Plants that have low and flexible shoots are preferable although many plants can be propagated using this method. Philodendrons, pothos, and syngoniums work well. For best results, layer woody growth from the previous season in early spring and the current season's growth in late summer.

1. **Preparation**- Prepare a growing space large enough to accommodate both the parent plant and the cutting. If you are layering outdoors, prepare the soil around the mother plant by adding sand, peat moss, or an alternative material to lighten the sand. The rooting medium should be able to retain moisture and air, yet also be able to drain well. If you are layering indoors, you have several options. To avoid transplanting, simply place your mother plant next to a pot of a similar height and lighten the soil of the mother plant with peat or similar medium. Alternatively, you can use a tray or pot which houses both the mother plant and the cutting. Although this set-up may be less cumbersome, it also adds potential stress due to several transplants.

2. **Selecting**- Choose a shoot which is low on the mother plant and fairly flexible. It needs to be able to bend without snapping.

3. **Wounding and Girdling**- Wound the stem by making a diagonal cut on the lower surface of the branch. The cut should not puncture more than half the branch and should be between nodes. You can hold the cut open with match sticks. Alternatively, girdle the branch by removing rings of bark about one inch wide. The wound or the girdle should be made on the portion of the stem which will get covered, about 5 to 10 inches from the tip of the stem.

4. **Rooting Hormones**- Rooting hormones can be applied to the wounded or girdled area to augment root formation.

5. **Layering**- Bend the branch down to the prepared rooting medium. Cover the wounded or girdled

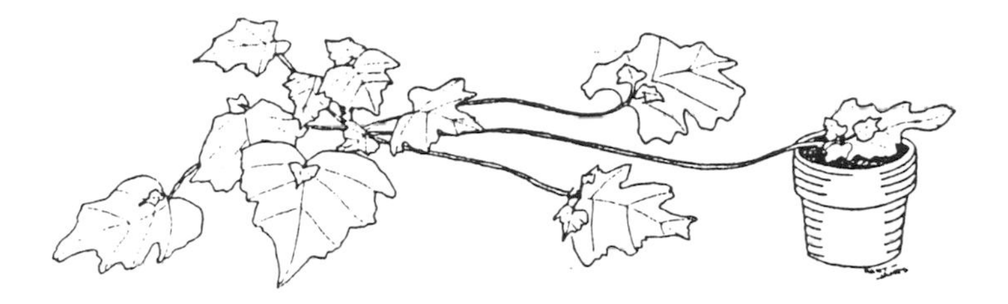

Layering

portion with rooting medium so that the top 5 to 10 inches of the tip is uncovered. You will probably need to anchor the layer with a stone or u-shaped wire pegs. Bobby pins work well. Roots will form on the covered, wounded area, so be sure that the soil or medium is packed well and in contact with the shoot.

6. **Separating**- Different species of plants root at different rates. For some difficult to root species, you will not be able to separate the cutting for up to two or three growing seasons. For other easy to root plants, you can separate the cutting in the same season. Gently pull on the layer to determine if it has developed a root system. If it has, cut the cutting away from its parent plant, making sure not to cut any of the new root system. If you have used two pots, you will not need to transplant immediately. When the root system is too large for the pot, transplant the cutting following normal transplanting procedures.

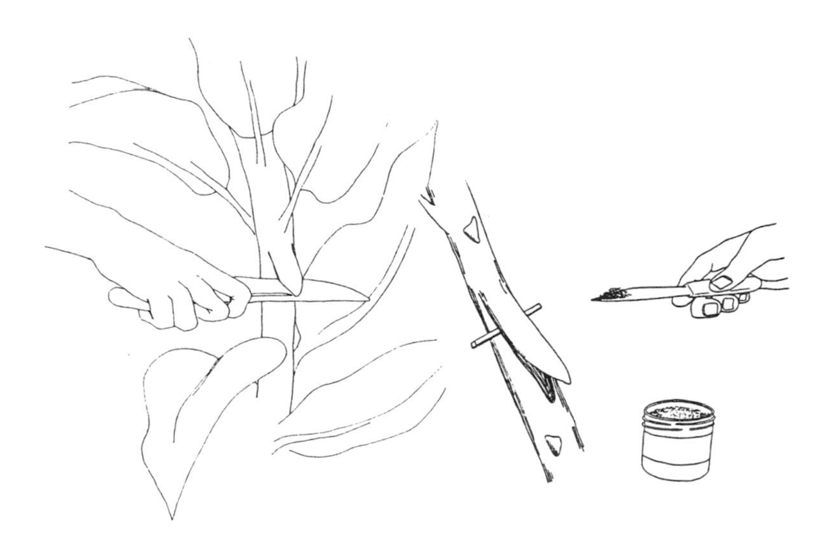

Air layering

Air Layering

Air layering works on the same principle as layering with one exception: you do not plant the wound in a growing medium. Instead, the wound is bandaged and roots form on the bandage, in the air. Air layering works because the wound traps carbohydrates that would have otherwise flowed down to the roots. Food is therefore provided to the area and water flows as usual because the cut is not deep enough to disrupt the flow of fluid through the xylem tissue which carries water. This method is excellent for plants which have become leggy. Not only are you producing a new plant, you are reshaping the old one. Be sure to choose an actively growing plant. Rubber plants are great candidates for propagation by air layering.

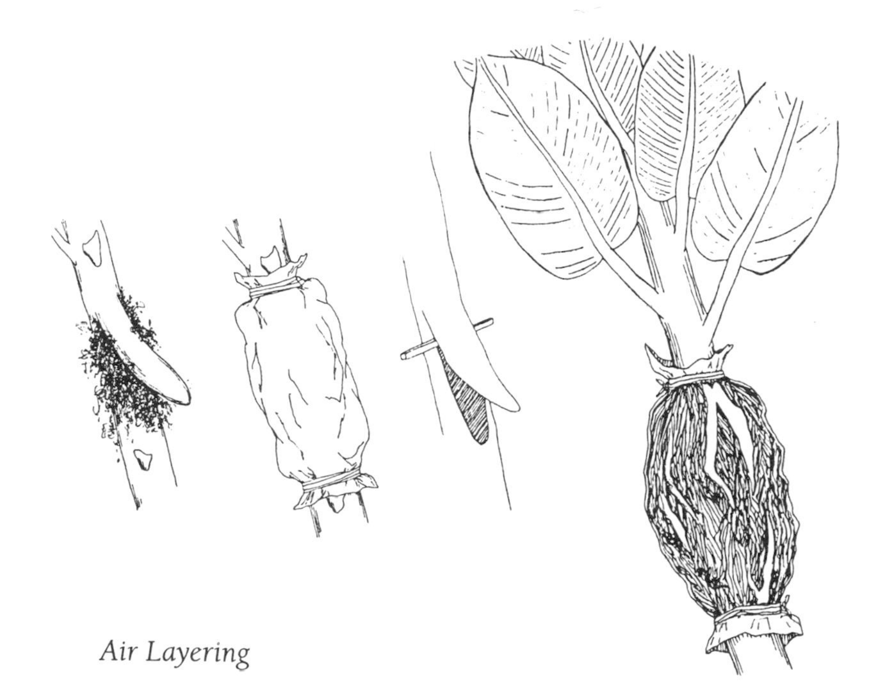

Air Layering

1. **Wounding or Girdling**- You want to wound or girdle the plant 10 to 12 inches from it's top. To girdle, use a sharp, sterile knife to remove a 1-1 1/2 inch band of bark encircling the branch. The girdle should be deep enough so that the underlying wood is exposed. Scrape the exposed wood to remove the cambium tissue (the thin actively growing layer directly under the bark). You can also wound the branch by making an upwards diagonal cut about 1 1/2 inches long. Insert a match stick in the wound to keep it from healing. This method may weaken the stem and eventually cause a break, so many propagators prefer girdling.

2. **Rooting Hormone**- Cover the wounded area with rooting hormone.

3. **Wrapping**- Soak peat or sphagnum moss for at

least an hour in water. Wring out excess water before wrapping the cut area with the moist peat or sphagnum moss. The moss should extend above and below the wound on the stem. You may have to remove several leaves to wrap the area sufficiently. Generally a six inch area is cleaned of leaves although the bandage will not be that large. Wrap the moss with plastic and secure the top and bottom. It is important that the bandage is secured firmly because it works to retain moisture and heat and aid in root formation. The plastic must extend slightly beyond the moss to achieve this. (If you are air layering outdoors use white plastic or aluminum foil so you do not trap too much heat.)

4. **Separation**- It will take several months for a strong root system to be visible. There is no guesswork with air layering. White roots will be clearly visible through the plastic when the plant is ready to be separated. Remove the plastic, leaving the moss, and cut the parent plant just below the new roots. Repot the new plant. The parent plant will begin to produce new leaves and will hopefully develop into a more bushy, well-proportioned plant.

5. **Aftercare**- The new plant will need to be watered generously, misted, and set in diffused light for several days.

Offsets

Some plants such as Spider plants, piggy-back plants and strawberry begonias produce tiny plantlets at their tips. If these tips are rooted, they will develop into new plants. There are several simple ways to achieve this.

Offsets

You can use the basic principles of layering and pin the offsets into small pots surrounding the parent plant. Use bobby pins of paper clips to pin the offsets down, making sure that the tips are well anchored and covered in rooting medium. Once the tip has developed strong roots, it can be severed from the mother plant.

You can also allow the parent plant to develop aerial roots on the shoots. These can simply be potted as small plantlets. Leave a small piece of stem from the parent plant on the shoot and make sure that it is well buried in the rooting medium. Create a warm, humid environment similar to that used for stem cuttings. A plastic tent works well if you do not have a humidity box handy. Keep the soil moist.

Cuttings in Water

Growing cuttings in water is perhaps the easiest means of propagation. It is a great project for children interested in gardening because it requires no equipment beyond that which is available in the average household. If you have fluorescent or HID grow lights, you can grow these cuttings under them, but it is not necessary. A sunny window sill will do. Ivies, Impatiens, Creeping charlies, Begonias, Spider plants, Geraniums, sweet potatoes, Coleus and many other varieties can all be propagated simply in a glass of water. To do this simply cut an offset or stem from a healthy parent plant and place it in clean, tepid water. Change the water every few days to provide a fresh oxygen supply and to avoid root and stem rot. If you forget to change the water, the roots will cluster near the top of the container in search of oxygen. This will stunt the plant's growth and promote algae formation.

A hydroponic method for rooting cuttings in water uses a small aquarium air pump, tubing, airstones or bubble wands. The stones or wands are placed in the bottom of the container and provide oxygen to the roots. This method works well with many varieties of plants; It is called the "sub-aeration" method.

Although it is fun to see the roots develop, this method works best if the tips are rooted in an opaque container to subdue or eliminate light. Sterilize an old mug in a mild bleach solution and use it. Once strong roots have developed, add a small amount of sand, soil, or rockwool granules to the water every few days. This will encourage the roots to grow new root hairs which provide the plant with food and water. It will also help you avoid having to untangle a mass of roots when it is time to transplant. When the root system is strong and imbedded in the medium, transplant the cutting to a larger pot.

Cuttings in water

Grafting

Grafting is a method of propagation whereby you cut two woody plants and encourage them to heal together as one. Grafting is done to strengthen a plant which has difficulty growing in its original form, but whose characteristics you like. Roses, for instance, are normally a grafted species, as are apple trees and numerous other plants. Nurseries combine varieties that have desirable flowers with varieties that are strong, disease-resistant, and easy to grow. Grafting is useful to the hobby gardener who has difficulty raising a particular variety of plant. There are many different ways to graft plants. Because grafting is most successful for shrubs and not very applicable to house plants, we will share the simplest way with you here. If you are interested in other methods, there are many good books which address grafting specifically.

1. **Selection-** Select two plants that are compatible and share a similar diameter. The more closely related the plants, the more likely that your graft will take. Choose a variety that is tough, disease-resistant, and healthy for the rootstock (the bottom part of the new grafted plant). The rootstock, also known as understock, should generally be one or two years old and 1/8 to 1/4 inches in diameter. The scion is the part that is grafted to the top of the rootstock. It should be a branch or stem with buds from which the desirable fruit or flower will grow. Choose a scion that is one years old and 4-6 inches long. It should be as close in diameter to the rootstock as possible. If the diameters differ, the scion should be smaller.

2. **Cutting-** Using an sharp knife (pocket, X-acto, kitchen, or grafting) cut the rootstock and the scion at the same angle, one upwards, and one downwards, so that they fit together neatly. Do not touch the cut part. The cut part must not be covered by rooting medium so that when the scion roots, it has not rooted into the medium.

3. **Binding-** If the two pieces are slightly different in diameter, line up an outside edge. Do not center the smaller piece, which is likely to be the scion, because the cambium layers need to connect in order for the graft to work. The cambium layer is between the bark and wood. It is a thin layer of actively growing tissue directly under the bark. If the cambium layers are not touching, your graft will not take. The cambium layers will not unite, even if they are touching, if they have dried out. Work quickly or store the scion and rootstock in moist sphagnum moss until you are ready to wrap. Tightly wrap the wound with thin wire, rubber bands which have been cut open, or tape. The wrap should hold the pieces firmly in place. Cover the wound with plastic in order to maintain optimum heat and humidity.

4. **Aftercare-** Keep the plant's growing medium moist, but not soggy. The surrounding environment should be warm and humid.

5. **Separating-** Within a few weeks, you will know if your graft succeeded. A callous will form at the wounded spot, and the branch will show growth. Leave the wrapping in place until the callous is pushing against it. When this happens, cut away the wrapping, being careful not to disturb the callous. Once the plant has healed, the grafted area will be its strongest.

Index

Gardening Indoors with CO_2

96 pages - illustrated - 5 1/2" x 8 1/2" - **$12.95**

Packed with the latest information about carbon dioxide enrichment - how to get the most out of CO_2 generators and emitters available today. Easy step-by-step instructions on setting up CO_2 in your garden room. Double your harvest with CO_2 .

Gardening Indoors with Cuttings

96 pages - illustrated - 5 1/2" x 8 1/2" - **$12.95**

Growing cuttings is fun and easy. This book is loaded with the most productive methods and information. Take cuttings to control plant growth and achieve super yields. Easy step-by-step instructions teach beginners and experts alike how-to take perfect cuttings.

Gardening Indoors with HID Lights

96 pages - illustrated - 5 1/2" x 8 1/2" - **$12.95**

This book is the definitive book on high intensity discharge (HID) lighting and plant growth. This book overflows with the latest information on high-tech lights. If you use HIDs, you must have this book.

Gardening Indoors with Rockwool

128 pages - illustrated - 5 1/2" x 8 1/2" - **$14.95**

New and updated volume of *Gardening: The Rockwool Book*. It is reformatted and packed with the latest information on rockwool.